Paper: The Continuous Thread

A THEMES IN ART BOOK

Published by The Cleveland Museum of Art
in cooperation with Indiana University Press

Paper

THE CONTINUOUS THREAD

Sheila Webb

An exhibition held at The Cleveland Museum of Art August 31 – October 24, 1982

Covers: Alan Shields, recto and verso of *Sun Moon Title Page* [30]

11150 East Boulevard, Cleveland, Ohio 44106

Distributed by Indiana University Press
Bloomington, Indiana 47405

Library of Congress Cataloging in Publication Data

Webb, Sheila, 1951-
Paper, the continuous thread.
(A Themes in Art book)
Bibliography: p.
1. Paper — Exhibitions. 2. Paper work — Exhibitions. I. Cleveland Museum of Art. II. Title. III. Series.
NK8552.5.U5C548 1982 702'.8 82-12840
ISBN 0-910386-69-2

Contents

Acknowledgements

The exhibition and accompanying catalog were made possible by the generous assistance of the Museum staff and the availability of the Museum's extensive collection.

I am grateful to Director Sherman E. Lee, who encouraged the original proposal, and to those members of the Department of Art History and Education who aided in various ways. I would especially like to thank the Curator, James A. Birch, for his overall supervision of the project; Associate Curator Helen O. Borowitz for her suggestions on the initial draft for this catalog; and Assistant Curator Marjorie Williams for her careful and sensitive editing of the draft. Janet L. Leonard, Assistant to Mr. Birch, not only typed the manuscript but also pointed out procedures that facilitated the exhibition process. Andrew Chakalis, Supervisor of the Extensions Division, and his staff shared my enthusiasm for this project and met any request — whether for an object to include in the exhibition or for information — with agreeable dispatch. Most of the contemporary paper pieces came from the Extensions Division.

Assistance from the Department of Prints and Drawings was invaluable. Chief Curator Louise S. Richards generously approved the loan of thirty-one objects. Secretary Robert J. Kurilla and Technical Assistant Charles G. Eiben helped with endless details. In particular, I want to thank Assistant Curator Jane Glaubinger, whose commitment to this project never flagged — from choosing works that illustrate particular aspects of paper, to discussing fine points of the manuscript, to checking credit lines for label copy, or in simply taking time with me to enjoy the selected work.

The Registrar, Delbert R. Gutridge, and his staff kept track of each loan. Associate Designer Joseph Finizia installed the exhibition to its best aesthetic and educational advantage; he also drew the map for the catalog. Museum Photographer Nicholas C. Hlobeczy and his assistants provided all of the photographs. Chief Curator Edward B. Henning and Curator Tom E. Hinson in the Department of Modern Art offered helpful suggestions. In the Publications Department, Chief Editor Merald E. Wrolstad's special care in the design of the catalog comes through in its elegant appearance. Associate Editor Sally W. Goodfellow edited the manuscript and expedited production details.

My thanks go to the artists who graciously sent their works — Walter and Mary Hamady, and Katherine Kuehn — and to an anonymous donor and the Dard Hunter Paper Museum, Appleton, Wisconsin. Leonard Schlosser, president of the Lindenmeyr Paper Company, New York, read the manuscript; his helpful comments are much appreciated, and only time constraints prevented incorporating more of them into the final copy.

Finally, I would like to thank my husband, Eric Friess, for his support and encouragement throughout this endeavor.

Sheila Webb
July 1982

Introduction

Paper is a humble substance with a venerable heritage. The ancient craft of papermaking, first practiced in China nearly two thousand years ago, embodies an honorable tradition in both the East and the West. Through the past centuries, the papermaker and the artist have been continually sensitive to both the manner in which paper was used and the methods of fabrication that determined the ultimate qualities of this practical yet highly creative material. As a continuous thread, paper unites artisan and artist, East and West, and traditional with contemporary art forms.

Necessity initiated the invention of this useful material. In China, the adoption of the animal-hair brush around 250 BC and the practice of calligraphy created a need for a cheap yet practical writing surface. Invented in AD 105, Oriental paper, made from vegetable fibers and formed on flexible bamboo molds, is an ideal surface to receive these tools. As the practice of papermaking followed the routes of the caravans West, each country adapted it to its own raw materials and specific needs. Paper made by the European mills was made from linen and cotton fibers, formed on a rigid mold and sized with gelatin. This technique stiffened the sheet and created a surface eminently suited to the quill and acidic ink made with oxgall and lampblack. Later, this same resilient sheet was adapted to the printers' press: type, the woodcut, and the engraving plate.

Until the twentieth century, paper played a supportive, secondary role in the arts. As a base for drawings, prints, and paintings it was a servant to artistic tools. In selecting paper as a support for their images, artists were familiar with and sensitive to the characterisitics of the individual paper sheet: its surface (texture, receptivity to ink, and level), color, and size. They emphasized the innate properties of paper in order to enhance their artistic statements.

It has been only during the last three decades that artists have recognized paper as an independent art form, a boldly assertive material that could be an active part of the creative process. The historian Dard Hunter (1883-1966) wrote extensively on the history and technique of papermaking; and the artist Douglass Howell (b. 1906), who established his own workshop in the 1950s, greatly influenced this awareness of paper as a material capable of artistic expression. These two men inspired the following generation of paper artists. Facilitating workshops established throughout the United States in the 1960s and 1970s provided an atmosphere conducive to experimentation and creativity. Ultimately, however, it was the attraction and appeal of the physical material itself that excited the imagination of contemporary artists. Some of these artists recall in their collages the many practical associations of paper: its communicative uses in books and advertisements, and even its utilitarian function as packing boxes. Others delight in the pulp itself — a pliable, sensual substance that may be molded, colored, shaped, dripped, or scarred in a seemingly endless variety of ways to create expressive forms and images.

The exhibition and accompanying catalog take as their theme the myriad aspects of this highly versatile material — paper — its history and methods of manufacture, its use in traditional art forms, and its elevation to an artistic medium.

The History and Techniques of Papermaking

Paper is properly defined as a substance made from fibers that are macerated until each individual filament becomes a separate unit. These fibers are mixed with water in a vat and then lifted from the water with a sieve-like screen. As the water drains, a sheet of matted fiber, or paper, remains on the screen. Any material that contains cellulose can be beaten and re-formed in this way. Paper is made the world over, since water and fiber are abundant, available, and inexpensive.

The invention of this flexible, malleable writing material is dated to AD 105 and is credited to Ts'ai Lun, an imperial guard in the Chinese emperor's court. Prior to this time, stone, bricks, metal, wood, leaves, bark, papyrus, silk, bamboo, parchment, and vellum were all surfaces upon which writing was executed. Paper, however, created different alternatives for literary and artistic expression due to its portability, reproducibility, and receptivity to ink.

Dard Hunter, the American scholar of papermaking, has postulated that the invention of paper resulted from the combination of two factors: the Chinese knowledge of feltmaking (the matting and intertwining of fibers)

and the desire to use waste scraps of silk trimmed from woven books and scrolls. He romantically suggests that workers took these scraps, beat them for re-use, and threw them on water to re-form. The appearance of "the minute fibers floating on the crest of water, as windblown seeds from the milkweed and dandelion float upon the surface of a pond or stream,"[1] inspired Ts'ai Lun to capture them on a screen and allow them to dry in the sun as a leaf or sheet. Ts'ai Lun's paper was made from the barks of trees, hemp waste, old rags, and fish nets. The knowledge of this process of making paper spread eastward — from China to Korea, and then to Japan — and westward — into central Asia and Persia via the well-traveled silk route, which extended from the Yellow Sea to the Mediterranean (Figure 1).

Oriental paper proved to be receptive to the calligraphic brush, and later, after the invention of printing, to the woodblock. In both the East and the West writing was practiced before paper existed, so most of the early paper was made specifically for the brush or quill. Oriental paper used for writing and printing is thinner, softer, and more pliable than European paper because it is made from vegetable fibers and formed on a flexible bamboo mold. Oriental calligraphy — writing characters with brush and ink — was perfectly suited to these thin, translucent sheets. Calligraphy is often explained through the age-old symbolism of feminine and masculine — the *yin* and the *yang*. The blank sheet of paper is thought of as *yin* — the receiver — and the brush as *yang* — the agent. The relationship between the two is a dynamic balance: "What is not *expressed* in ink is *suggested* in blank paper."[2]

The paper used for calligraphy and painting was adapted to the art of woodblock printing. In the earliest instance of printing in Japan (AD 770), only one side of the translucent sheet was used. Even today, this procedure of printing on only one side is followed.

Figure 1. The chronology of hand papermaking.

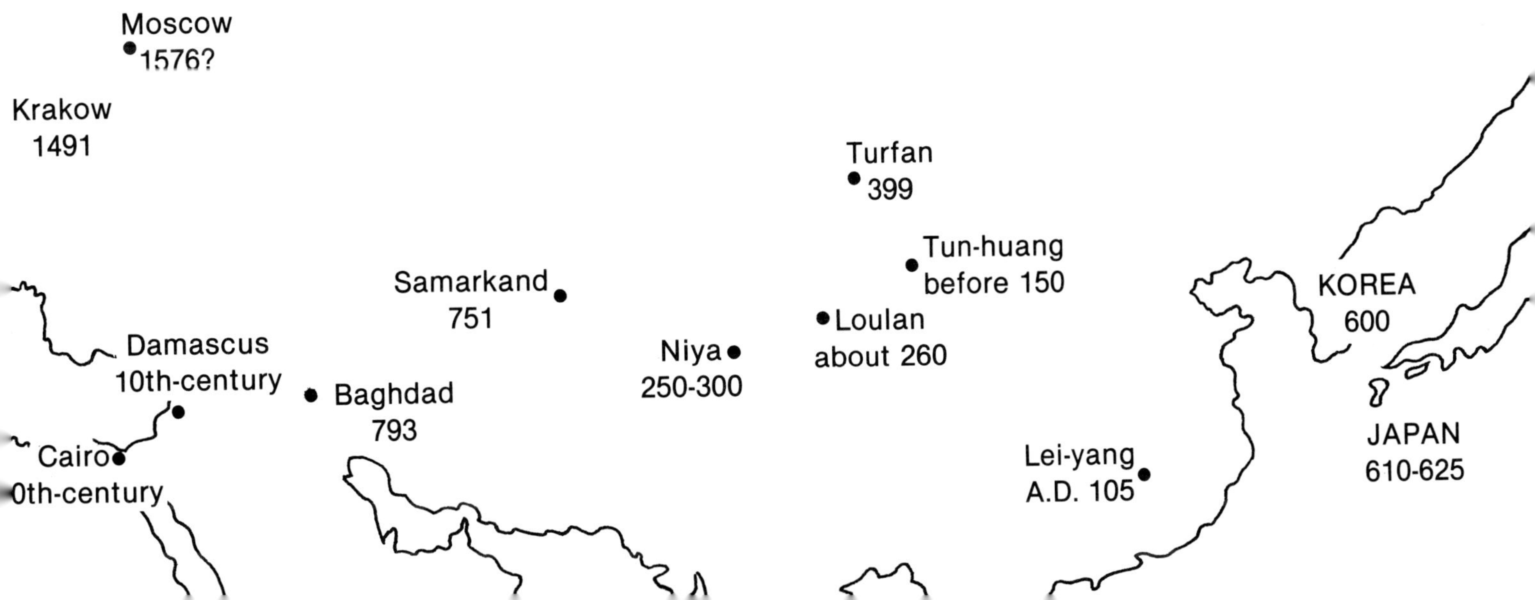

Paper is revered in Oriental cultures in a way unknown in the West. Immediately after its invention, paper coins replaced metal coins buried in tombs: "'Cunning knaves' cut paper into forms and offered it to the spirits in lieu of genuine silver and copper coins."[3] After his trip to China during the thirteenth century, Marco Polo described in his writings the burning of fragile paper effigies as ceremonial offerings. This practice both antedated his visit and continues presently as a hope that the departing spirit will possess, in afterlife, those things represented by the paper objects that, when burned, resume their original form in heaven. These may take the form of shoes, hats, robes, or even full-sized carts, horses, and automobiles.[4] In Japan, white paper folded and cut in streamers consecrates all places where it is hung: rocks, trees, and shrines.[5] Paper is thought to be an expression in itself — it occupies a place where the inner world and external statements meet.[6]

In its westward travels, paper may have been available in Samarkand before the eighth century, but it was not manufactured there until AD 751. An old Arabic manuscript, *Roots of Trades and Kingdoms,* tells of two Chinese papermakers who were taken as prisoners during a raid on Chinese territory by the governor of Samarkand. The captured craftsmen offered to impart their knowledge of making paper — the fabrication of which had been, like silk, closely guarded by the country of its origin — in exchange for their freedom.[7] Samarkand emerged as a major papermaking center because of abundant supplies of pure water and plentiful harvests of flax and hemp. From there the craft spread to northern Africa — Baghdad, Damascus, Egypt, and Morocco — and finally to Europe in the twelfth century. Artisans in each country adapted papermaking to their native resources — flax, hemp, or linen. From its introduction in Egypt, paper seems to have been accepted quickly. A letter of thanks written around 884 closes with: "Pardon the papyrus," an apparent apology for not making use of the new writing material.[8]

The cross-continental journey of paper took 500 years. Upon its arrival in Europe, however, paper was accepted at a "snail-like pace".[9] The guarded response to the new substance is generally explained by the fact that paper was thought to be perishable and expensive to produce. It was also associated with Judeo-Arabic origins, which were suspect to Christian Europe.[10] Scribes hesitated to give up their "beloved parchment," and aristocratic patrons who commissioned the scribes and illuminators also preferred the old methods.[11] Even after paper mills became common, important legal documents continued to be recorded on parchment — as are diplomas today, for instance.

Johann Gutenberg's invention of typography by 1450 created an increased need for paper. With this need established, raw materials for the mills in the form of cheap linen rags became available as fashion changed and linen underwear replaced the woolen variety. The mills then became dependent on rag dealers for raw materials. In fact, laws were eventually passed to protect the mills and keep them in supply. In fifteenth-century Genoa, for example, anyone not belonging to the rag-pickers' trade was forbidden to buy or sell used linen.[12] With establishment of overseas colonies, cotton came into use first as clothing and then as rags for papermaking. Around 1860 wood pulp replaced cotton rag, and it is used today in commercial mills.

Technique: East

> "Why is *washi* [Japanese paper] so wholesome? When we try to figure it out, we cannot help but think it is because nature is paper's mother and tradition paper's father."[13]
>
> Soetsu Yanagi, co-founder of the Japan Folk Art Society

The method of papermaking invented in China nearly two thousand years ago is similar to the technique used today in both East and West. The raw fibers, tools, and results are, however, quite different. Oriental paper can substitute for glass, leather, rubber, cloth — indeed, for almost any material. Paper can be thin and translucent, to be used as windows, providing light, warmth, and privacy. Paper can also be strong enough to make flooring, as is commonly done in Korea, where homes are heated by flues under the floor.

Paper artisans in Japan appreciate the raw materials used to make paper for their individual characteristics or "personality." *Kozo, gampi,* and *mitsumata* are the three most commonly used fibers. *Kozo,* called the masculine member of this vegetable trio, is made from mulberry inner bark and prized for its strength. Layers of *kozo* paper were used to line the breastplates of medieval Japanese warriors. This fiber was also used during the twentieth century to make balloons that transported bombs across the Pacific Ocean during World War II. Nine thousand balloons, each about forty feet in diameter, were launched from three sites near Tokyo. Timing devices were set to explode the bombs some forty to fifty hours later. An unknown number of balloons did reach the American continent; one journeyed as far east as Michigan.[14]

Gampi fiber is used for manufacturing papers in which ink absorption is not desired — for calligraphy, for making fans, or for tracing paper. It is sometimes formed on a screen covered with a silk cloth, which creates a smooth surface without the lines from the mold visible in other Oriental paper. *Mitsumata* is considered the feminine member of the Japanese fibers. It is "soft-complexioned" and rather absorbent, and *mitsumata* paper is perfect for stationery and the *fusuma* screens in Japanese homes. Although all of the plants used for papermaking grew wild, today *gampi* is the only one that has not been domesticated.

In Japan, where papermaking is traditionally a family or community concern, the basic elements needed in the papermaking process are found: paper-fiber trees and cold, soft water. Planting the paper-fiber trees is the first step of the long process. After harvesting, the tree branches are steamed to loosen the fibers of the inner bark. The fiber is then peeled, dried, soaked, treaded upon, scraped, rinsed, and re-dried. This turns the dark "black bark" to light "white bark." In the seemingly endless process, the fibers are further boiled, rinsed, and bleached. Bleaching may be accomplished naturally by leaving the fibers in the sun or in a running stream; or it may be done chemically. Papermakers poetically describe the bunches of fibers floating in dark streams as "white stars suspended in deep-valley rivers." After bleaching, the fibers are again beaten to break apart the individual strands to shorten, soften, and hydrate them — so that a tight, consistent sheet is possible. Today, machines have replaced the wooden handbeaters. The hand method crushes and separates the fibers, thereby preserving their strength. In contrast, the machine cuts the fibers. These clean-cut ends have been likened to the "hair of a cat" (Figure 2).

The forming of individual sheets begins only after the pulp is ready. A two-by-four-foot vat, or "papermaker's boat," is filled three-quarters full of water and balls of pulp are thrown in with the mucilage, most often called *neri* (starch). *Neri,* extracted from the roots or bark of trees, is comparable to gelatin sizing in Western paper. It looks like egg white and serves to make the fibers float evenly in the vat; it also keeps the fibers from becoming entangled when the individual sheet is formed. A mechanical agitator mixes the solution of pulp and *neri* to an even consistency.

The papermaker plunges the mold, a flexible bamboo screen placed on a frame made of bamboo and silk threads, into the vat. The papermaker scoops the solution up with the "vatman's stroke" and tosses the excess pulp off the back edge of the mold. Artisans compare this toss, which takes from two to four years to perfect, to the "waves of the sea." The papermaker rocks the mold back and forth, moving the solution over and off the mold until the sheet is formed by the "discharge" method. The sound of the pulp sloshing over the edge of the mold also resembles a wave. "That particular sound represents the solution at the right temperature and with the correct amount of *neri* moved across the screen by the experienced papermaker in tune with the rhythm" of the craft.[15]

As each sheet is made, it is put, or couched, onto the next sheet to form a stack, or post. This pile of individual, wet sheets of paper is then pressed for a number of hours to remove most of the water. In Europe, each sheet is separated from the next with a felt. In Japan, however, the sheets are stacked on each other; the *neri* prevents them from sticking during pressing. The sheets are then peeled from each other and brushed onto wooden boards to dry outdoors in the sun, or indoors over a heater. The side facing the board is the side used for printing. The papers are then trimmed and packaged for sale. In this way a papermaker in Japan can make one thousand sheets a day.

Figure 2. Steps in papermaking, from a Japanese woodcut, 1798.

In all, papermaking is a cold, difficult process. Traditionally, paper was made during the winter months when the artisans were not occupied with farming or other activities. A low temperature in the workshop is essential to avoid mold, as the plant fibers are still living materials even after cutting. The cold water and air also tighten the plant fibers, producing a crisp, fresh sheet. The beating and treading was formerly done barefoot in cold, running streams. Today paper can be made in any month of the year, since preservatives to retard spoilage are now available,[16] but many hours of difficult — though rewarding — labor are required. The attitude of the Japanese papermaker is identical to the approach of the Japanese artisan toward all crafts: "an admiration for simplicity and rusticity," while at the same time "there is a feeling for virtuoso use of materials and extreme refinement of technique."[17]

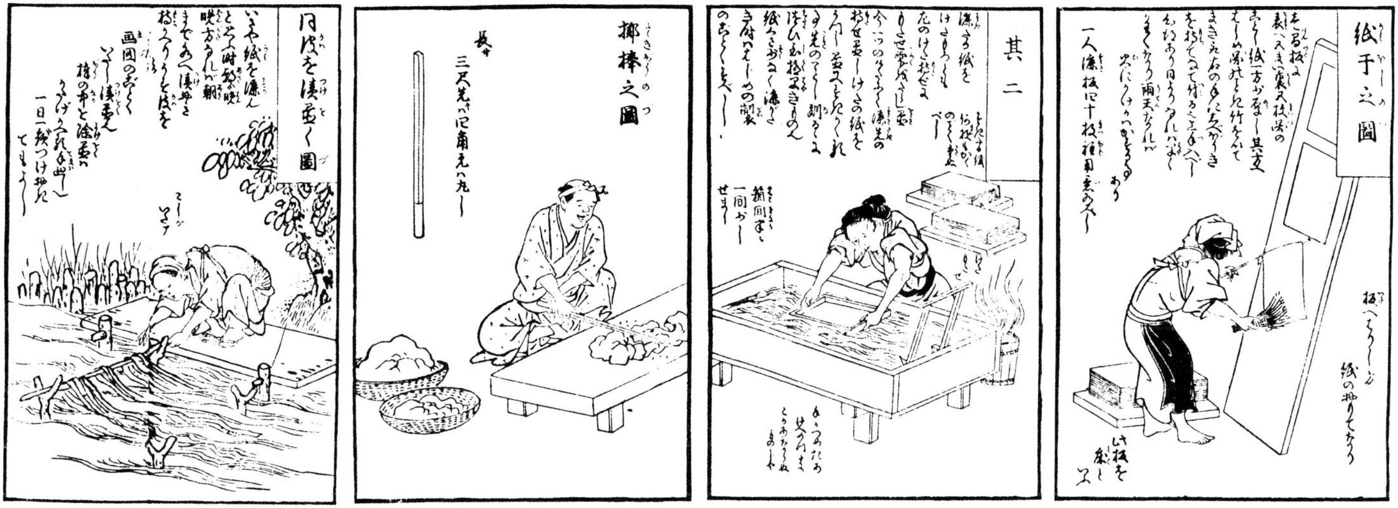

Technique: West

> "Heaven does not permit such a divine art to be made easy for mortals here below."[18]
>
> Fr. Imberdis, *Papyrus sive Ars conficeindae Papyri,* 1693

Paper in the Western world is considered practical and utilitarian. The materials used in its manufacture — initially, unyielding linen plus sizing, now wood pulp — limit and define its use as a surface to receive pencil or ink.

As in the East, the process begins with the collecting and preparing of available cellulose material, which was at first hemp and linen, and later, cotton. The Arabian practice of retting the rags was adopted by European papermakers. Rags from well-worn cloths were wetted, pressed into balls, placed in piles, and allowed to ferment.[19] Like a compost heap, in six weeks to two months the heat generated by the pile of material breaks down the fibers. As much as a third of the rags were lost, but the beautiful buff color characteristic of early papers results from this procedure. The rags were then boiled in water with wood ash and suspended in bags in a running stream to clean them.

Next, the fibers were beaten. Even at the very beginning of papermaking this process was mechanized. In a seventeenth-century woodcut (Figure 3) water-powered stampers can be seen in the background. The beaters worked like a mortar and pestle to break down the rags. A machine invented in the Netherlands by 1680 replaced the stamping mill. Called the Hollander, it consists of an oval vat partially divided into halves. A motorized cylinder fitted with blades rotates against a stationary base plate also complete with blades. The vat is filled with water and the rags are added. The rags are gradually refined into pulp as they are propelled around the vat.

After the fibers are completely hydrated into pulp, the "stuff" is put into a vat of water. A "hog," or mechanical agitator, is often used to mix the water and pulp. The mold used in European papermaking is made of wire on a rigid wooden frame. A wooden deckle, or fence, is

Figure 3. Elias Porcelius, *Curionsen Spiegels (Strange Mirror)*, leaf 15. Published by J. A. Endter, Nuremberg, 1689. Engraving from the woodcut lent by the Dard Hunter Paper Museum, Appleton, Wisconsin.

placed on the mold to help form and hold the sheet. After placing the deckle on the mold, the vatman dips the mold into the vat, lifts it up, and allows the water to drain on the "tripan," the platform that extends across the vat. The coucher removes the deckle; places the mold on its side; and gradually lays it face down on a felt, transferring the wet sheet onto it. The coucher passes the mold back to the vatman, and the process begins again.

Paper is classified as "laid" or "wove" depending on the type of mold used to form it. In a laid mold, lines of metal wire are stretched across a wooden frame. These laid lines are formed in the wet pulp and their texture can easily be seen, especially if you hold a sheet up to the light. George Seurat's drawing of 1887, *Café Concert* (see [8]), clearly shows the "hills" and "valleys" formed by such a mold. Around 1750, the printer John Baskerville invented the "wove" mold. He wanted to create a smooth surface for his delicate type faces (the printing surfaces of movable type). The wires for the molds were woven on a loom in the same manner as cloth. When the sheet is formed it has a surface like woven silk without a distinctive pattern of horizontal and vertical lines. This smooth surface proved to be eminently suitable for lithography, invented in 1798.

The deckle also leaves an impression on the sheet along its edge where the pulp squeezes out; this is called the "deckle edge." A machine method has been invented to duplicate this effect that was originally a mark of the artisan's hand.

The watermark is a final characteristic of the paper sheet that depends solely on the mold. Artisans may have begun using the watermark as a means of identifying the mold size, as a trademark, or merely for fun. The watermark is made of wire and is fashioned into the shape of animals, human figures, plants, names, or other symbols. Laid on the face of the mold, it leaves an imprint on the wet pulp. In Seurat's *Café Concert* the watermark "Michalette" can be clearly seen on the left side of the drawing. It may be read vertically from top to bottom. As Seurat drew on the surface of the paper, the conté crayon made visible the texture and design created in the paper by the watermark.

After a certain number of sheets have been formed and couched, the stack, or post, is pressed. Traditionally, a platen was put on top of the post and was then placed in a screw press. A bell rang, calling all the workers in the mill to operate the screw press. Papermakers today use a hydraulic press to remove most of the water from the sheets. Pressing removes eighty percent of the water and reduces a two-foot post to a six-inch post. Each sheet is then peeled from its felt and hung in sets of five on ropes strung from beams in drying lofts. Contemporary hand papermakers dry their sheets between a sandwich of blotters, cardboard, and particle board. Fans or drying chambers hasten the drying process.

After drying, the sheets may be dipped in animal sizing. Writing papers require more sizing than those intended for printing. Papermakers called the sizing room the "slaughterhouse" since so many sheets were lost in this process. The sheets were next burnished by hand with stone, a practice adopted from the Orient. Burnishing closes the pores of the paper and renders it more suitable for writing with ink. When mechanized glazing hammers were brought into use in the early seventeenth century, labor disputes erupted in the industry due to the opposition of the hand burnishers. By 1720 sheets were also calendered, or pressed, between wooden rollers to finish and smooth them.

The finish of an individual sheet has three designations: "rough" paper is pressed only once; "cold-pressed" paper receives a second pressing without felts; "hot-pressed" paper is fed through zinc plates without heat to further smooth out the pores. Additional finishing procedures more radically alter the sheet: it may be filled with ash and then coated with clay to create the continuous refined surface necessary for fine-screen halftone reproduction (as in offset lithography).

As was true in the Orient, the lives of Western papermakers were difficult. They worked twelve hours, from 6:00 a.m. to 6:00 p.m., and ate their lunch while working. The stooping, the warmth of the stock into which they plunged their hands, and the lack of light and heat in the mills all created a wet, arduous occupation — yet one in which they took great pride. Although most paper is now commercially produced by means of more sophisticated methods and machinery, contemporary hand papermakers still follow the old methods.

1. Dard Hunter, *Papermaking, The History and Technique of an Ancient Craft* (New York: Dover Publications, 1978), p. 77.
2. Sukey Hughes, *Washi, The World of Japanese Paper* (Tokyo, New York & San Francisco: Kodansha International, 1978), pp. 159-60.
3. Hunter, *Papermaking*, p. 203.
4. Ibid., pp. 207-8.
5. Hughes, *Washi*, p. 25.
6. Ibid.
7. *Papermaking, Art and Craft* [exhibition catalog] (Washington: Library of Congress, 1968), p. 16.
8. Hunter, *Papermaking*, p. 470.
9. Jules Heller, *Papermaking* (New York: Watson-Guptill, 1978), p. 187.
10. Ibid.
11. Edward Lucie-Smith, *The Story of Craft, The Craftsman's Role in Society* (Ithaca, New York: Cornell/Phaidon Books, 1981), p. 139.
12. André Blum, *On the Origins of Paper*, trans. H. M. Lydenberg (New York: R. R. Bowker, 1934), p. 37.
13. As quoted in Hughes, *Washi*, p. 150.
14. Hunter, *Papermaking*, pp. 220-21.
15. Hughes, *Washi*, p. 100.
16. Ibid., p. 74.
17. Lucie-Smith, *Story of Craft*, p. 82.
18. As quoted in Hunter, *Papermaking*, p. 196.
19. Ibid., p. 153.

The Characteristics of Paper in Relation to the Artistic Image

The choice and the use of paper by the graphic artist are important elements in determining the final effect of a print or drawing. Each of the objects selected for this exhibition in some way illustrates the qualities and characteristics of paper. For purposes of discussion, the selection has been organized around three factors inherent in the sheet: 1) surface (texture, receptivity to ink, and level); 2) color; and 3) size.[1] In any print or drawing none of these elements is mutually exclusive, because each relates to the others in a harmonious and satisfying way.

Surface

A. Texture and Receptivity to Ink

All paper has a texture determined by the type of fiber, the type of mold used, the kind and amount of sizing, and the finishing. Historically, there has been a tendency to use a textured, heavier sheet in the relief processes — letterpress and woodcuts — and a smoother sheet for intaglio and lithography. In the relief process all nonprinting areas are carved, engraved, or etched away. The lines and areas left on the orginal plane surface are inked and printed by hand or in a press. In intaglio, the printing areas are recessed. The plate is inked and then wiped clean, leaving ink in the depressed areas. The ink is transferred to damp paper by running plate and paper through an etching press. Lithography is a planographic (flat writing) process in which prints are made from a flat surface rather than from incised or relief areas of a plate. A design is made on a stone with a greasy crayon. After the stone is dampened, ink is rolled across its surface. The ink, which sticks to the greasy crayon marks and not to the wet stone, is transferred to paper in a press.

In general, the woodcut image tends to be rougher and bolder, whereas the intaglio and the lithographic images often have finer linear detail and more subtle tonal areas. Each process enhances the texture of the sheet. In the relief process the paper touches only the inked areas, thereby revealing the texture of the paper in the negative space. The intaglio or lithographic processes, however, further smooth and press the sheet due to pressure from the plate or stone during printing.

Paper's surface texture and its receptivity to ink are closely related. Receptivity, or the way paper absorbs ink, depends on several factors: 1) the softness and absorbency of the fibers; 2) the method and length of maceration (the longer the maceration, the shorter the fibers, and the more brittle the sheet); 3) the type of mold; 4) the kind and amount of sizing; 5) the degree of finishing (polishing, calendering, coating, etc.). In printing, paper's receptivity to ink is naturally dependent upon the actual printing process when such factors as the pliancy of the sheet and its ability to respond to the inked plate affect the final image. Controls such as dampening the sheet and altering the amount of pressure applied during printing also influence the paper's willingness to accept the image.

Before paper was available, parchment and the more refined vellum made from the skin of cows, sheep, or goats, were the materials used for writing and printing in Europe. Paper's ability to absorb ink was in sharp contrast to vellum. The vellum surface is hard and repels ink, paint, and graphite. When printing on vellum the ink is not absorbed but rather is smashed from the plate onto the surface and spreads out across it. The ink can then lift or flake off, which destroys the image. The fifteenth-century anonymous German woodcut on vellum, *Madonna and Child with Three Members of the St. Augustine Order* [1], can be compared to *Schatzbehälter* [2], a woodcut on paper by Michel Wolgemut (1434?-1519). In the first print the smooth, shiny surface of the vellum is obvious, and the ink has rubbed off. In the latter, the relief lines of the woodblock are clearly impressed in the sheet of the paper. The negative or white areas stand out to reveal the laid texture of the sheet.

The *Florilegium*, or flower book, dating from 1608, shows how the impervious surface of the vellum receives paint. The images on the page illustrated here [3] retain the crisp, clearly delineated strokes of the brush. The paint lies on the surface rather than being absorbed into the material. Consequently, the paint can actually be erased, as is evident on the worn lower left margin of the page.

Like paint, pencil and ink also ride or sit on the surface of vellum. *Louis XIV Captured by Venus* [4], a seventeenth-century work from the studio of Charles Le Brun, exemplifies the effect of pencil and ink on vellum. The shiny surface of the vellum is apparent in the outlined areas. The pencil marks, especially in the less detailed background, stand out crisp and smooth, since

1 Anonymous, *Madonna and Child with Three Members of the St. Augustine Order*, ca. 1473.

2 Michel Wolgemut, *Schatzbehälter (Reliquary)*, 15th century.

there is no surface texture to interrupt or absorb the graphite.

Historians estimate that it took the skins of 300 sheep to provide enough parchment to print one Gutenberg Bible.[2] This single statistic illustrates the importance of paper to the graphic arts. First of all, numerous editions of texts and prints were possible. Paper, created from available and more plentiful materials, was about one-sixth the cost of parchment (even though it was still considered an expensive item).[3] Secondly, its ability to respond to ink was critical to the print media. Typographic printing, invented around 1445 by Johann Gutenberg, was meant to work best on the hard, gelatin-sized paper made by the European mills. Early bookmakers continued the form of illuminated manuscripts in both layout and type design. A page from a sixteenth-century book [5] shows type printed with a woodcut illustration. The letterforms and image could be printed at the same time in the press. The combination of paper, produced from pure water and rags, and ink, which gave a crisp impression on the sheet, and printing the sheet while damp to save the typeface and aid in absorption, all created a sensuous object unlike a modern, commercially produced book.

The feel and look of fine relief printing on handmade paper has kept the craft of handset letterpress printing alive. In England in the last third of the nineteenth century, William Morris (1834-1896) was instrumental in starting the Arts and Crafts movement, which aimed to reunite the artist and the product in the Industrial Age. He established the Kelmscott Press and published fifty-three titles. In addition, he learned to make paper by hand, and he studied medieval manuscripts and the incunabula (books printed before 1501). His book and type designs derived from these early sources (see [39]).

Morris's revival also influenced the Roycrofters, an

3 Anonymous, *Florilegium* (folio 42), early 17th century.

arts and crafts group in the United States. Dard Hunter, who spent seven years at their shop in East Aurora, New York, as art director, loved the handmade paper used to print books — so much so that he dedicated his life to documenting the paper trade around the world. Between 1903 and 1910 he designed over one hundred titles. He also published a book entirely by his own hand: he made the paper, cast the type, and did the press work and the binding. Almost singlehandedly, he kept alive the interest in and knowledge of handmade paper.

4 Studio of Le Brun, *Louis XIV Captured by Venus*, 17th century.

5 Hans Wolff, *Illustration from a Printed Book,* 16th century.

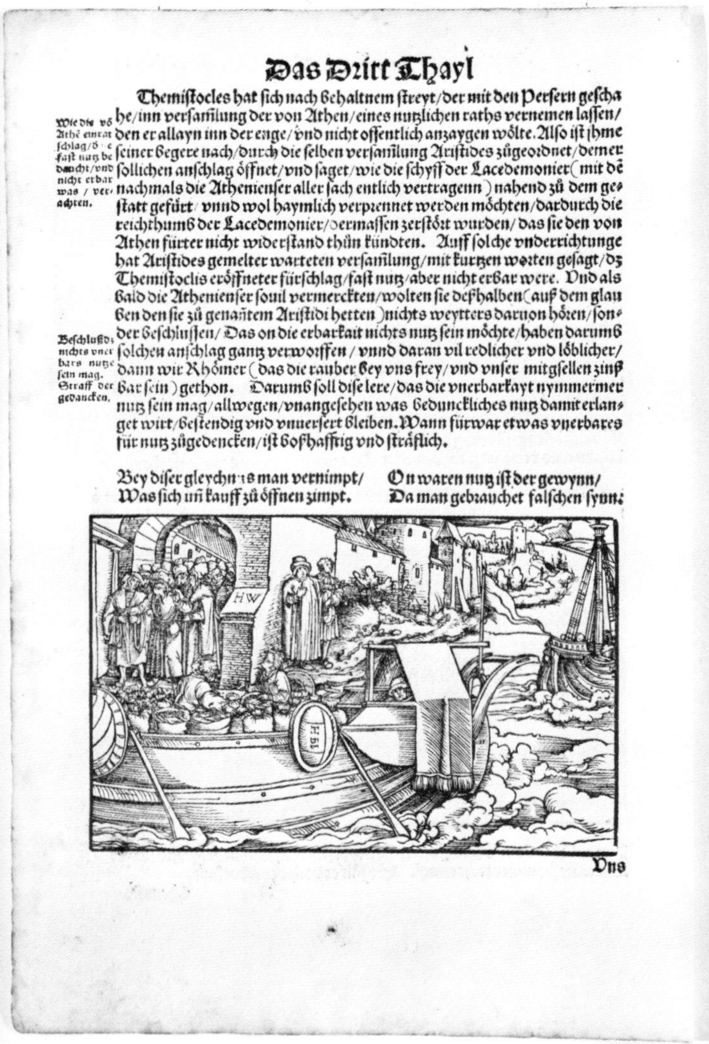

Das Dritt Thayl

Themistocles hat sich nach behaltnem streyt/der mit den Persern gescha he/inn versamlung der von Athen/eines nutzlichen raths vernemen lassen/ den er allayn inn der enge/vnd nicht offentlich anzaygen wölte. Also ist jhme seiner begere nach/durch die selben versamlung Aristides zügeordnet/demer sollichen anschlag öffnet/vnd saget/wie die schyff der Lacedemonier (mit dē nachmals die Athenienser aller sach entlich vertragenn) nahend zü dem ge-statt gefürt vnnd wol haymlich verprennet werden möchten/dardurch die reichthumb der Lacedemonier/dermassen zerstört wurden/das sie den von Athen fürter nicht widerstand thün kündten. Auff solche vnderrichtunge hat Aristides gemelter warteten versamlung/mit kurtzen worten gesagt/dz Themistoclis eröffneter fürschlag/fast nutz/aber nicht erbar were. Vnd als bald die Athenienser souil vermerckten/wolten sie deßhalben (auf dem glau ben den sie zü genañtem Aristidi hetten) nichts weytters daruon hören/son-der beschlussen/Das on die erbarkait nichts nutz sein möchte/haben darumb solchen anschlag gantz verworffen / vnnd daran vil redlicher vnd löblicher/ dann wir Rhömer (das die rauber bey vns frey/vnd vnser mitgsellen zinß bar sein) gethon. Darumb soll dise lere/das die vnerbarkayt nymmermer nutz sein mag/allwegen/vnangesehen was bedunckliches nutz damit erlan-get wirt/bestendig vnd vnuersert bleiben. Wann fürwar etwas vnerbares für nutz zügedencken/ist boßhafftig vnd sträflich.

Wie die vō Athē einrat schlag/d[illegible] fast nutz be dacht/vnd nicht erbar was / verachten.

Beschluß d; nichts vnerbars nutze sein mag. Straff der gedancken.

Bey diser gleychnüs man vernimpt/
Was sich uñ kauff zü öffnen zimpt.

On waren nutz ist der gewynn/
Da man gebrauchet falschen synn:

Vns

In the last twenty years there has been an increasing interest by artists in the combined use of paper and the letterpress. The graphics department at the University of Wisconsin in Madison, Wisconsin, headed by Walter Hamady is a major center in this movement. Walter Hamady (b. 1940) and his wife, Mary (b. 1948), also run the Perishable Press and have so far published 105 titles. *Papermaking by Hand* [6 & 7], one of their most recent books (written by Walter), is a step-by-step guide to this "white art." Each signature (section of a book made from a folded sheet) contains different kinds of paper. Some are Shadwell, the Hamady brand, while others are European (such as Fabriano, from the oldest mill in Europe) or Oriental. In studying the book, the reader becomes aware of the great variety in papers, the way each responds to the type's impression, the way each page receives ink, and the relationship of ink color to paper. Hamady, in the illustration by Jim Lee [6], is shown at the vat forming a paper sheet; his lively description of the procedure appears on the facing page.

Special sheets bound into the back of Hamady's book demonstrate particular papermaking techniques. The sheet illustrated [7] contains imprints from string, letterforms, and geometric shapes. These were laid on the sheet after it was couched, or placed, on a felt. After the sheet was pressed and peeled from the felt, it retained the impression of these objects. The facing gray sheet exhibits partially embedded strips of paper. Diluted colored pulp — purple, pink, and blue — added to the surface creates a painterly effect. Contemporary bookmakers such as the Hamadys are reviving a 500-year-old process that achieves a unity of image and technique.

One of the joys of printing by hand is the way in which the texture of the paper influences the final image. The surface texture also influences the drawn

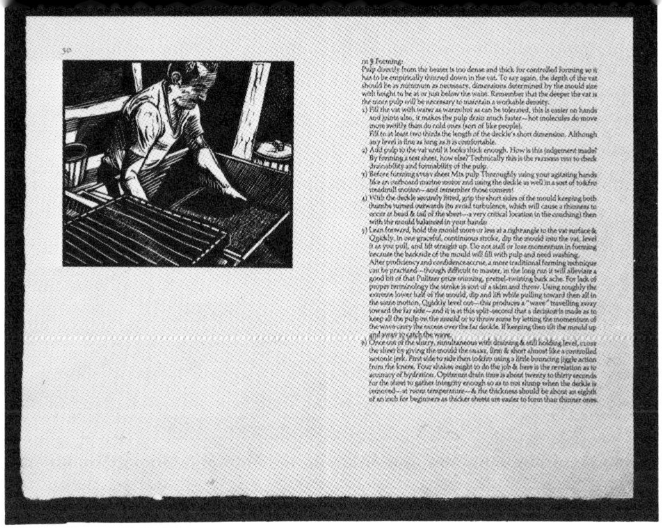

30

III § Forming:
Pulp directly from the beater is too dense and thick for controlled forming so it has to be empirically thinned down in the vat. To say again, the depth of the vat should be as minimum as necessary, dimensions determined by the mould size with height to be at or just below the waist. Remember that the deeper the vat is the more pulp will be necessary to maintain a workable density.

1) Fill the vat with water as warm/hot as can be tolerated, this is easier on hands and joints also, it makes the pulp drain much faster—hot molecules do move more swiftly than do cold ones (sort of like people).
Fill to at least two thirds the length of the deckle's short dimension. Although any level is fine as long as it is comfortable.
2) Add pulp to the vat until it looks thick enough. How is this judgement made? By forming a test sheet, how else? Technically this is the FREENESS TEST to check drainability and formability of the pulp.
3) Before forming EVERY sheet Mix pulp Thoroughly using your agitating hands like an outboard marine motor and using the deckle as well in a sort of to&fro treadmill motion—and remember those corners!
4) With the deckle securely fitted, grip the short sides of the mould keeping both thumbs turned outwards (to avoid turbulence, which will cause a thinness to occur at head & tail of the sheet—a very critical location in the couching) then with the mould balanced in your hands:
5) Lean forward, hold the mould more or less at a right angle to the vat surface & Quickly, in one graceful, continuous stroke, dip the mould into the vat, level it as you pull, and lift straight up. Do not stall or lose momentum in forming because the backside of the mould will fill with pulp and need washing.
After proficiency and confidence accrue, a more traditional forming technique can be practised—though difficult to master, in the long run it will alleviate a good bit of that Pulitzer prize winning, pretzel-twisting back ache. For lack of proper terminology the stroke is sort of a skim and throw. Using roughly the extreme lower half of the mould, dip and lift while pulling toward then all in the same motion, Quickly level out—this produces a "wave" travelling away toward the far side—and it is at this split-second that a decision is made as to keep all the pulp on the mould or to throw some by letting the momentum of the wave carry the excess over the far deckle. If keeping then tilt the mould up and away to catch the wave.
6) Once out of the slurry, simultaneous with draining & still holding level, CLOSE the sheet by giving the mould the SHAKE, firm & short almost like a controlled isotonic jerk. First side to side then to&fro using a little bouncing jiggle action from the knees. Four shakes ought to do the job & here is the revelation as to accuracy of hydration. Optimum drain time is about twenty to thirty seconds for the sheet to gather integrity enough so as to not slump when the deckle is removed—at room temperature—& the thickness should be about an eighth of an inch for beginners as thicker sheets are easier to form than thinner ones.

6 Walter Hamady, pages from *Papermaking by Hand*, 1982.

7 Hamady, pages from *Papermaking by Hand*.

line. In *Café Concert* [8] by Georges Seurat (1859-1891), laid lines of the sheet have virtually become part of the composition. As Seurat's crayon moved across the sheet it emphasized the ridges in the paper's surface. The roughness of the sheet and the contrast of the rich black conté crayon against the white paper create the sense of a darkened hall with spotlights illuminating only the stage. Seurat rendered spectators and performer as geometric masses of dark and light, and the grid evident in the sheet increases the sense of geometry. As the best-known practitioner of Pointillism (color applied in small dots to a surface so that from a distance they blend together), Seurat achieved a similar effect in this drawing. The "valleys" in the paper that the crayon passed over stand out as little dots of light and lend even the blackest areas a shimmering luminosity.

The contemporary artist Mel Pekarsky (b. 1934) also exploits surface texture in his work. His series of drawings illustrating the American Southwest — which includes the work from 1979, *Funeral Mountains* [9] — are done on rough-textured paper normally used for water color. The surface of the paper itself suggests the variegated topography of the desert; the pencil scraping across this surface furthers the illusion of shimmering heat, scrubby brush, and jutting rocks. The large areas left white and untouched create the impression of infinite space.

The surface texture of paper has actually provoked the development of some artistic techniques. Soft-ground etching is a print medium invented at the end of the eighteenth century specifically to reproduce the quality of a drawing, since collectors during this period wanted to own drawings. Soft-ground etchings were initially

8 Georges Seurat, *Café Concert*, ca. 1887.

9 Detail.

9 Mel Pekarsky, *Funeral Mountains – Study #5*, 1979.

done in red-brown ink to replicate the sanguine drawings popular at the time. *Seated Bather* [10], done in 1905 by Pierre Auguste Renoir (1841-1919), recalls the rococo style in color and in subject.

To make a soft-ground etching, the metal plate is prepared with a ground that does not completely harden and therefore remains sensitive to pressure. Textures pressed onto the ground make a pattern which can then be etched into the plate. Renoir placed a sheet of laid paper on top of the ground and executed the drawing. Where pressure was applied, the ground adhered to the underside of the sheet. When the sheet bearing Renoir's drawing was lifted off the plate, it pulled away the ground, leaving a negative area in the plate that was etched. Even the texture of the laid lines of the sheet

was transferred onto the ground and subsequently to the plate.

The laid sheet registers not only its own texture but also, as in any drawing, influences the crayon's mark on the sheet. *Seated Bather,* actually printed in red-brown ink on wove paper, has the quality of a crayon drawing on a laid surface similar to Seurat's *Café Concert.* The plate mark, where the plate pressed into the paper during printing, is our clue that this is not a crayon drawing.

Special paper used to transfer a drawing to a lithograph stone may register its texture in a similar manner. Although artists can, of course, draw directly on the stone, they may for various reasons prefer to work on paper. It is easier to carry a piece of paper into the field than to transport a heavy stone. James McNeill Whistler, for example, worked on transfer paper in the French countryside and then mailed the drawings to his London publisher, who transferred them to the stone. If an artist drew on highly textured paper, then both the texture of the sheet and the image would be transferred onto the lithograph stone. Transfer paper, handled like a page from a sketchbook, allows the artist's hand great freedom. Käthe Kollwitz (1867-1945) produced her haunting *Self-Portrait* [11] in this way. The stippled line evident in the print has the quality of crayon on paper rather than ink on stone. The uninked white paper represents dappled light, which forms abstract patterns across the face, emphasizing the serious expression.

While artists such as Seurat and Renoir utilize the texture of the paper's surface as an essential element in their compositions, other artists strive to eliminate the texture in the sheet. Certain techniques perform better

10 Pierre Auguste Renoir, *Baigneuse assise (Seated Bather),* ca. 1905.

on a smooth wove sheet. One critic has suggested that lithography might not have been invented were it not for the invention of wove paper about 1750, which preceded Alois Senefelder's discovery of lithography in 1798. In his lithography manual of 1818, Senefelder recommended using only smooth, lightly sized paper.[4] Because lithography is a planographic process, the paper receives ink from only the surface of the stone. A smoother sheet is more apt to pick up and retain all of the detail present on the stone. Lithography allows the ink to float on the sheet and is different from relief processes that emphasize a three-dimensional impression.

Don Eddy (b. 1944), in his lithograph *BMW Showroom* [12], solves the technical problems inherent in creating illusions of reality and spatial depth on a flat surface. Reflections from the high-gloss surfaces — glass, chrome, and lights — are ingeniously combined with reflected depth. The cars inside the showroom are superimposed on the cars parked alongside the curb in the street outside. Trees and buildings are reflected in the windshield of the white car, and fluorescent room lights float inside the building façade, which is reflected from across the street. Lithography allows for the almost-photographic accuracy of the flat image, which gives a sense of depth while it piles up information.

The refined paper surface is even more critical in such commercial printing processes as offset lithography and the half-tone screen process. Major advances in the production of highly finished paper made these techniques feasible. As early as 1822 Michael Faraday tested the ingredients of India paper, which was known to be successful in printing illustrations. He found, through burning a sheet and weighing the remaining ash, that the paper contained a filler. This filler, by closing up the pores in the sheet, had a direct effect on the paper's surface.[5]

11 Käthe Kollwitz, *Self-Portrait*, 1915.

After the invention of photography in 1839, methods were sought to reproduce the photographic image in traditional print form for use in books and newspapers. This "reproductive revolution" culminated in the 1880s with the half-tone screen process.[6] The half-tone screen breaks down an image into a dot pattern that reproduces a varying range of grays onto a printing plate. The plate can then be printed in a press with type. A surface that accepted the ink with no apparent interference was essential: the highly refined surface was made possible by first smoothing the sheet through calendering (smoothing between rollers) and secondly, by using a filler to close up the interstices between fibers. Finally, the surface of the sheet is coated with clay adhered with casein adhesive.

Robert Rauschenberg's *Sunday News Magazine Cover* [13], done for the *Miami Herald,* demonstrates the use of mechanized printing methods in contemporary art. Rauschenberg (b. 1925) breaks down the distinction between lithography as a fine printer's medium and its use in mass media by collaging negatives, bits of ads, technical manuals, and letterforms. The photographic reproductions in this catalog were also printed by using the same half-tone screen process. A magnifying glass held up to *Sunday News Magazine Cover* will reveal the dot pattern.

The illusion of reality at which photography excels is, of course, enhanced if there is no obvious surface texture to distract the eye. Invented by Niepce and Daguerre, the daguerreotype was made on a sensitized, highly polished metal plate and was available by 1839. People enjoyed examining the minute details made possible by the camera obscura with a magnifying glass. The English photographer William Henry Fox Talbot's rival process, also available by 1839, used sensitized paper to make both negative and positive. The paper's texture, evident in the photographic image, was disliked by the public, who complained of lines visible in the print. These lines were thought to be especially disturbing in portraits. Because of this and the long exposure time required, initially Talbot's method was used mostly for landscape scenes. Later adaptations of Talbot's process used coatings such as albumen for photographic paper that were smooth and glossy and ideally suited to rendering any detail or tonal range present in the negative onto the print. Documentary photographers capitalize on the medium's ability to capture and retain detail and tones. By printing even the subtlest tones from the negatives, modern photographic paper can give the appearance of looking through a window onto a scene.

With the availability of photography by 1840, artists working in other media were inevitably influenced by the new process and experimented with it. Between 1851 and 1853 members of the Barbizon School in France experimented with etchings produced using photographic methods. Jean Baptiste Camille Corot (1796-1875) and Charles François Daubigny (1817-1878) were the best-known practitioners of *cliché-verre.* A glass plate was coated with collodion and the image scratched into it with a needle or other tool. A black cloth could be placed behind the glass to aid in seeing, and therefore executing, the design. This plate was then put on a light-sensitive piece of paper and exposed to sunlight. Where the collodion had been scratched out, the paper turned black. The plate Daubigny used to make *A Cluster of Alders* [14] did not extend to the edge of the paper and therefore created a black border. The edge of the image is slightly fogged where light seeped in under the plate.

The effect is very different from that of a similar image etched in a plate and printed on paper. The quality of

12 Don Eddy, *BMW Showroom,* 1973.

13 Robert Rauschenberg, *Sunday News Magazine Cover,* 1979.

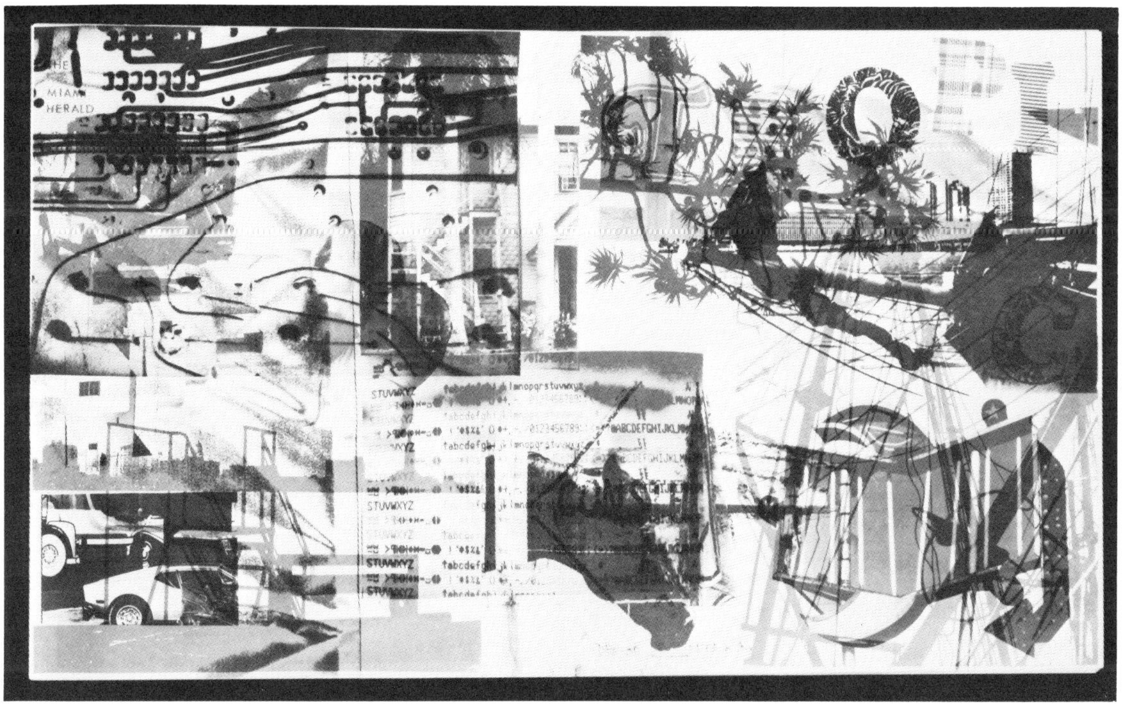

the line of a *cliché-verre* can be compared to ink on vellum: smooth and glossy. Since the paper is coated with albumen, the same substance used at the time in printing negatives, it has a smooth, shiny surface. This effect, which is due not to the paper but to the coating, has been called "harsh" and is one reason printmakers, accustomed to the texture of paper, abandoned the procedure.[7]

In contrast, some photographers devised specific procedures to avoid the smooth photographic surface. In an attempt to align themselves with the "higher arts," photographers created methods at the turn of the cen-

14 Charles François Daubigny, *A Cluster of Alders,* 1862.

15 Clarence White, *Julia Hall McCune,* 1903.

tury that used photographs as only a starting point toward an image that more closely resembled an etching or painting. Highly textured paper was coated with platinum or pigmented gum bichromate. The negative to be contact printed onto this prepared paper might then be scratched to resemble the etched lines of an intaglio plate. Sometimes the final image was handcolored.

Clarence White (1871-1925) was a member of the Pictorialist School, a group of photographers who used these techniques. His portrait of *Julia Hall McCune* [15], done in 1903, has the soft, diffused romantic quality characteristic of the period. The female figure in a domestic setting was a favorite theme for White. In this print the paper was sensitized only in selected areas. The swirling motion of the brush used to apply platinum to the sheet is evident, especially at the perimeter of the image. The photograph reflects the quality of a sketch; this quality is heightened by the pencil strokes that follow brush marks along the chair legs and the folds of the dress and even stand alone to complete the hem. Today, contemporary photographers are reviving these earlier techniques that eliminate the smooth, commercially produced surface.

The characteristics of modern photographic paper include texture, surface gloss, thickness, and tint. Each step in producing the photographic paper influences the final image, just as textured or smooth paper affects the drawn or printed image. Photographers choose the paper suited to their requirements. Paper with a smooth surface reveals finer detail and any apparent grain in a negative. Paper with a textured surface is frequently chosen for printing portraits. The surface sheen of photo paper — glossy, lustre, or matte — also influences the print. A glossy, smooth surface will show a greater tonal range — dark and light areas on a very glossy surface will appear more brilliant than on a matte surface; it also

16 Jerry Uelsmann, *Untitled,* 1964.

17 Robert Trostle, *Geometrics I – Roller Coaster, Mexico City,* 1978.

renders more detail and grain in the print. The rough or matte surface scatters light, decreasing the tonal range.[8]

A photographer such as Jerry Uelsmann (b. 1934) exploits our willingness to believe in the illusion of reality — at which the photographic image excels. By combining negatives he creatives a dream world of intense vision. His iconography includes folded hands, growing and decaying plants and trees, and peapods — all suggesting a primordial cycle of growth and decay. Although each section of his photograph [16] from 1964 looks realistic, the total composition (printed from multiple negatives) is a created world of trees, sky, and leaf combined in a unique way. Trees line the horizon against a white sky. Two-thirds of the frame is filled with a wispy gray, veined leaf. The central vein of the leaf melts into the tree trunk above it. The brilliantly printed images combined in a surreal way create an image of great tension and beauty.[9]

Contemporary color materials dispense with the paper surface entirely. Composed of paper coated with plastic, they greatly increase the high-gloss surface of paper-based photography. The two illustrated examples use imagery that is perfectly suited to this plastic surface — a roller coaster in Mexico City by Robert Trostle (b. 1944) and the close-up of a pinball machine by Charles Bell (b. 1935). The slick appearances of the images are further enhanced by the garish color and shiny surfaces of the subject matter.

Geometric I – Roller Coaster, Mexico City [17] is visually composed of the three primary colors: red distinguishes the horizontal bars, while blue paints diagonals, and yellow colors the verticals. Only the green palms in the lower right corner add a fourth color to the piece. The intricate structure was photographed so that it completely fills the frame, creating an abstract pattern of linear color.

Pin Ball [18] was also composed to completely fill the frame with color. The metal pinball reflects and curves the room behind the photographer. The sheen on the ball is also apparent on the highly polished surface of the machine. Red letters whiz around a blue and yellow background; lightning bolts seem to lift off the surface like a rocket. In both photographs, the plastic shiny surfaces and saturated colors increase the sense of abandon and play.

Paper's receptivity to ink and its influence on a printed image is clearly seen in *Figure I* [19] by Hilda Volkin (b. 1933) and *Dune Landscape* [20] by the German Expressionist Ernst Ludwig Kirchner (1880-1938). The artist's deliberate use of less absorbent paper is evident in Volkin's *Figure I.* The Oriental wove paper has no apparent texture in the areas of negative space, which are seen as the smooth skin of the body and the white lines of patterning on the couch. The non-absorbent paper holds the lines delineating the hair, the blanket, and the diagonals of the couch on the surface so that they stand out crisply. The edge of the inked area that outlines the figure, as well as the small dots that delineate the spine, buttocks, leg, and toes are also distinct. The ink, which appears to be a subtle gray, is actually bands of green and blue. Although visually they blend together, upon close examination the bands can be discerned as separate: each color is a band that overlaps its neighbor. The recumbent woman, the flattened treatment of her white form against the background of solid gray, and the decorative patterns of the blanket and couch all combine to create an image reminscent of Japanese prints.

In contrast, Kirchner consciously selected a highly absorbent paper for his depiction of *Dune Landscape.* His choice is a reflection of the late nineteenth-century interest in materials as both creators of an image and substances in themselves that influenced the German Expressionists. These painters and printmakers working during the early years of the twentieth century strove to make their artistic techniques and creative processes obvious. (The example most often cited is the Expressionists' use of highly absorbent blotting paper in printing woodcuts.) They aggressively exploited the woodcut medium, one of the oldest methods of relief printing, through the use of heavily gouged lines. In *Dune Landscape* wide black strokes alternate with areas of the block left uncarved, which allowed the wood grain to print its natural impression onto the paper. Kirchner's choice of blotting paper as a support for his work heightens the

18 Charles Bell, *Pin Ball,* 1981. See also color plate.

primitive quality of the image. Pressure during printing forces the soft paper sheet toward the lower levels of the carved block, where it can pick up the slightest deposit of ink. The absorbent and pliable blotting sheet may be used to add mottling, to model forms, and to print gray areas. The crisscross pattern characteristic of blotting paper — evident in large black areas — recalls again the nature of the materials.

19 Hilda Volkin, *Figure I*, 1973/74.

20 Ernst Ludwig Kirchner, *Dune Landscape with a Curving Bay*, 1913.

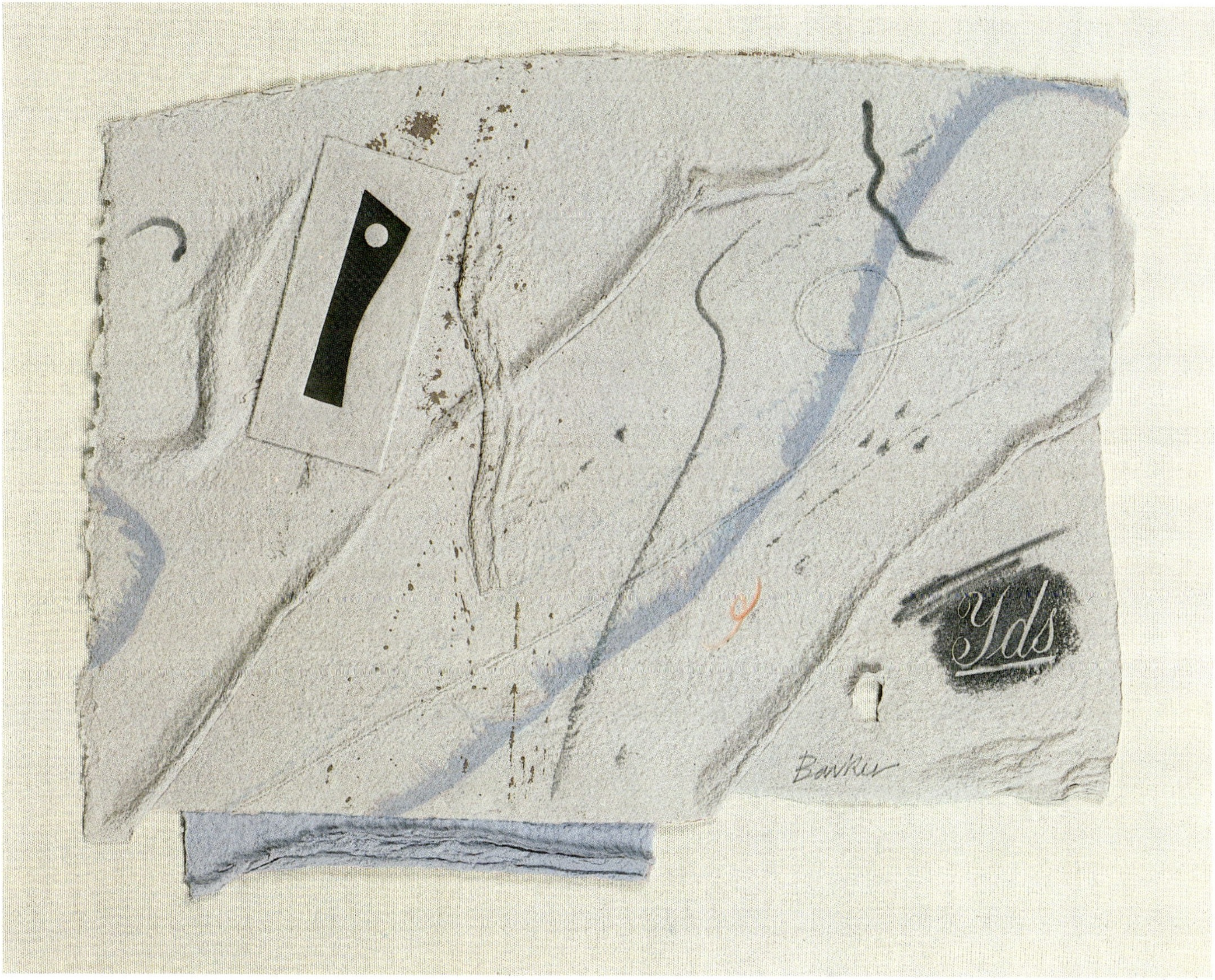

50 Laurence Barker, *Untitled #5*.

29 Ed Moses, *Wedge 5*, 1973.

28 Frank Stella, *Untitled,* 1967.

18 Charles Bell, *Pin Ball,* 1981.

B. Level

Level is a term used to describe the three-dimensional quality evident on the paper's surface. The two most common ways of introducing level to the sheet are through printing and embossment. The varying levels of the sheet may be coordinated with, or independent of, inked areas.

Claire Falkenstein (birthdate unknown) exploited the ability of paper to be molded three dimensionally in her contemporary etching *Untitled* [21]. Two embossed circles interlock in the upper half of the composition. Energetic etched lines move around and out of the circles like little tadpoles or calligraphic marks. The shapes of the dots and squiggles are repeated along the perimeter of the sheet by using an etching plate with an irregular edge. The three-dimensional quality of the embossed circles is recalled in the impression caused by the edge of the plate. An irregular horizontal greenish-yellow line introduces the one element of color and anchors the composition.

Using a reserved vocabulary, Victor Vasarely (b. 1908) exhibits great versatility in the embossed screenprint of 1967, *G-FF* [22]. Through level and color, *G-FF* plays with the way the eye perceives depth. The shadows cast by embossed geometric shapes add to the composition. In a subtle manipulation of form, the ink is limited to circles and squares, only some of which are embossed. The pattern of inking sometimes echoes an embossed shape, an inked circle within an embossed circle; at other times it does not. The vertical bands at either side of the design — one of diamonds within squares and the other of circles within squares — are uninked. The diamonds and circles not only reverse their level but also their direction. The horizontal bands that border the image at top and bottom reverse the position of their inked areas; only the diamonds reverse their level. The precedence our eye gives to lighter colors is reversed:

21 Claire Falkenstein, *Untitled*, 1969.

22 Victor Vasarely, *G-FF*, 1967.

some white areas remain uninked and recede in depth. Black, a color that normally recedes, here pops out. This sophisticated handling of depth and color lends vibrancy to *G-FF*.

Extase [23], done in 1967 by Paolo Boni (b. 1926) and illustrated here with the plate from which it was printed [24], demonstrates sculptural embossing. The plate is made of metal shapes on which selected portions have been etched. The paper retains the impression and configuration of the metal shapes pressed into it. The crudely etched hatchmarks; the inking that separates each shape from its neighbor; and the imprint in the sheet from nails used to rivet the metal forms to their wood base recall the hand processes used in constructing the plate.

In relief printing, the lines of the woodcut or the typeface are impressed into the surface of the sheet. This can cause the white, or uninked areas, to stand out. Like other German Expressionists, Emil Nolde (1867-1956) used the relief process in a bold manner for stark effect, as seen here in *Dancers* [25]. Some areas of the woodblock were cleaned of ink before printing. This process models large masses of white areas, as is evident in the folds of the dancers' dresses.

The artist can layer, cut, tear, or even weave paper to give level to the sheet. The contemporary artist Louise Nevelson (b. 1899) in her collage *Beyond-Q17* [26], first drew in black ink on white paper. A torn piece of black paper was pasted on top of the drawing. The black ink, accented by the black paper, stands out sharply against

23 Paolo Boni, *Extase*, 1967.

24 Boni, *Extase*, etching plate.

25 Emil Nolde, detail of *Dancers (Tänzerinnen),* 1917.

the white background. The black L-shaped paper with two uneven edges is massive compared to the delicate, undulating line. Tension is created between the two levels through the contrast between the clean, torn edges of the black paper and the freely drawn thin line on the white paper.

Number 16 ("Visual Delights" Series) [27] by another contemporary artist, Ellen Stavitsky (b. 1931), is a tight pack of multiple layers of found paper accentuated with silver and pink thread. Each of the torn fragments is differently colored; the colors range from muted blues to pale silvers. Much of the piece is hidden or only suggested. Although print is visible on the third layer of paper, its message or meaning is obscured by both the sheets layered on top and by the silver paint that covers it entirely. As Stavitsky fixed each piece in the "Visual Delights" Series on a blank sheet of paper, it became an exploration of the "different means of dividing an empty space."[10]

In contrast to works that demonstrate level due to the physical manipulation of the paper's surface, the lithograph *Untitled* [28], by Frank Stella (b. 1936), achieves level through color. The flat image recedes or advances due to the placement of colors and the decreasing size of the squares. Four triangles in each of the two major squares are outlined in pencil, accenting the strong vectors intersecting the center. The graph-paper grid was printed first, followed by a separate plate for each color represented in the geometric formation. The grid is either obscured or revealed in selected rows of color by the transparency of opacity of the ink. The lines of color have a soft, flowing edge as if fashioned with a watercolor brush. They are in sharp contrast to the rigid lines

26 Louise Nevelson, *Beyond-Q17*, 1977.

28 Frank Stella, *Untitled,* 1967.
See also color plate.

27 Ellen Stavitsky, *Number 16 (''Visual Delights'' Series),* 1979.

of the graph. Pencil lines added by hand follow the rigidity of the graph to outline the squares. Other pencilled lines dissect the squares, to accentuate the triangles formed through color. These pencil lines were freely drawn so that, like the rows of color, they overflow the boundaries of the rigid graph. The varying patterns of colors within the squares and the complex relationships between the eight triangles establish a strong sense of forward and backward motion. The eye perceives the flat geometric forms as three-dimensional.

Wedge 5 [29], a lithograph by Ed Moses (b. 1926), combines three sheets of transparent tissue paper each printed with varying intensities of colored ink to create an illusion of a misty depth. The thin tissue allows col-

29 Ed Moses, *Wedge 5,* 1973. See also color plate.

ors on the two underlying sheets to show through. The ink seems to float on the surface of the smooth tissue. Moses's choice of simple, muted pastels corresponds to the delicacy and fragility of the paper. Both the paper and ink work in unison to suggest hidden images viewed through soft netting. The thick and thin orange lines of the bottom layer form diagonals, while yellow, blue, and orange blotches of color float like water spots on the middle layer. The gray diagonals on the top layer are less brilliant than the underlayers of color and are not continuous lines; they break, only to start up again. This muted gray complements the veiled intensity of the colors underneath. All three layers combine to give the appearance of streamers or confetti crisscrossing and overlapping themselves in a vertical rush down the sheet.

Sun Moon Title Page [30], by Alan Shields (b. 1944), uniquely designed to be viewed from both sides, combines traditional techniques of screenprinting, monoprinting, and watercolor with sewing, collage, and weaving. Each side is united to the other physically and visually. The pattern of gray checks on one side (verso) is recalled in the stamped forms on the other. The weaving motif becomes three-dimensional where the paper sheet was slit and watercolored strips woven in and out of both sides of the piece. Machine-sewn thread affixes triangles to one side of the sheet (recto) and wanders in and out of the greenish half-moons on the reverse side.

Level is apparent not only in the collaged, woven, and threaded areas but also is achieved through inking. On the two triangular, collaged pieces, watercolor is absorbed and swims out over the paper (recto). Stamped shapes that resemble celestial bodies are applied next. Since they overlap in places, the eye can trace the progression in which the layers have been applied. Thread is the last layer and final touch of color.

Color

All prints and drawings have color, based on the tones of ink and paper used. "White" is the term used for a wide variety of light tones. The range of white tones may vary, depending upon the location of the paper manufacturer, the pulp, the purity of the water, and dyeing or bleaching. The tone of the paper may change later due to mold, exposure to chemicals, sunburn from exposure to light, or even stains from handling.

The buff color of papers dating from the fifteenth to seventeenth centuries is due to the yellowing of rags during the fermentation process. By the eighteenth century an increasing brightness appeared in paper, due in part to the introduction of cotton fibers. Cotton, a whiter and softer cloth than linen, is easier to decompose and does not need fermentation. The invention of the Hollander beater by 1680 also eliminated the need for retting (allowing rags to rot in piles). Since the mechanical

30 Alan Shields, *Sun Moon Title Page* (recto and verso), 1971. See also front and back covers.

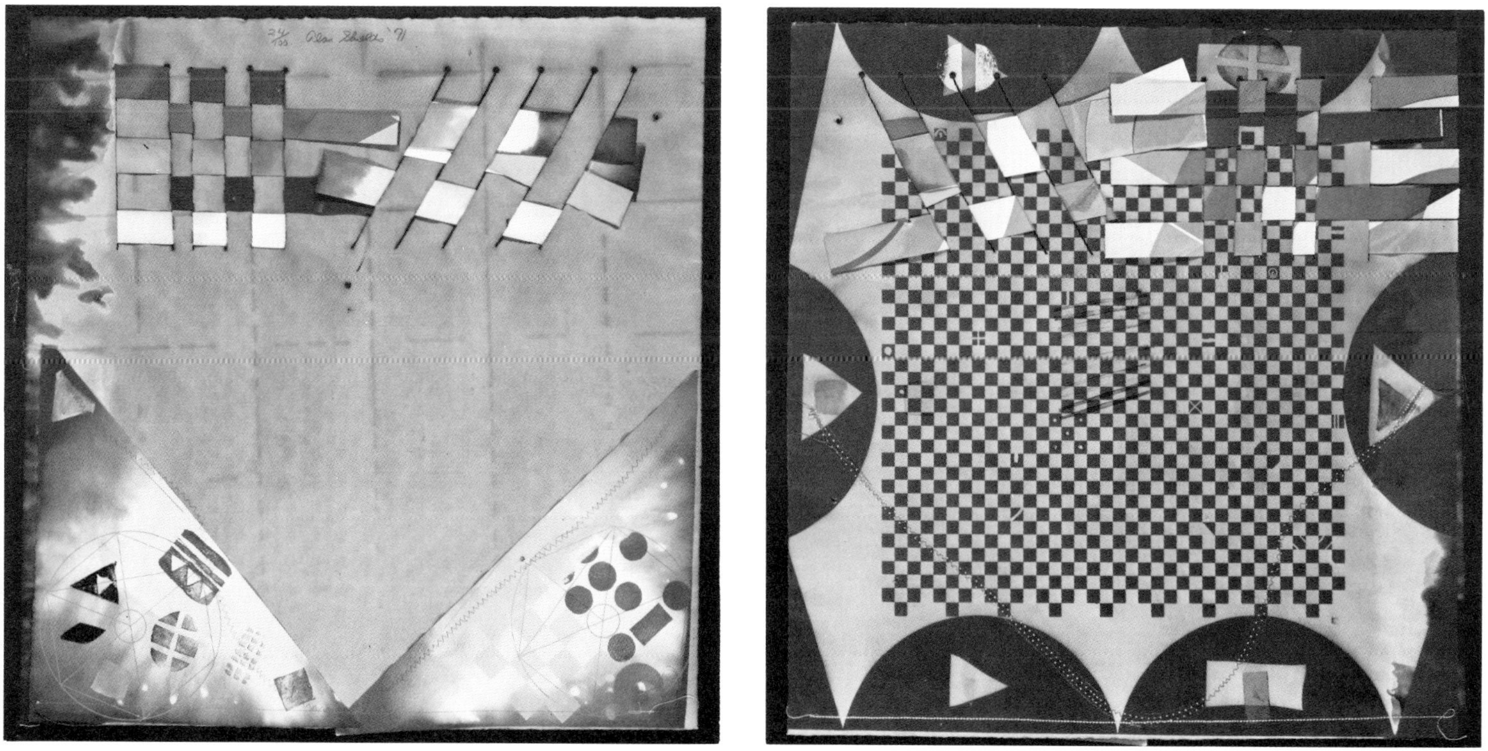

blades of the Hollander broke down the rags quickly and efficiently, the Hollander was readily adopted in mills across Europe.

Early printmakers viewed the white of the paper merely as a blank background for their images. In the early woodblock prints the white areas were frequently handcolored. Artists gradually recognized the aesthetic qualities of these white tones and integrated them into their final image. Andrew Robison, curator of prints and drawings at the National Gallery of Art, credits the early printmaker known as the Master of the Playing Cards with the first use of white in a positive role — that of hatching to show qualities of light and shadow. Hatching, a technique of fine, closely set parallel lines, is used to shade or model and may "portray light falling to lessening degrees"[11] around curved forms, thereby giving a figure three-dimensionality. The use of this technique to indicate volume and light is evident in the *Schatzbehälter* (see [2]) from the fifteenth century and in a more developed manner in the sixteenth-century *Illustration from a Printed Book* (see [5]).

The Black Hat [31], a lithograph portrait of his wife, Emma, by George Bellows (1882-1925), demonstrates the influence that color can have on a print. It is printed on two different sheets of Oriental paper — pure white and creamy white — only one of which is seen here. The black crayon marks shading the face and dress stand out crisply against the white sheet. In contrast, the creamy paper softens the overall effect, imbuing Emma with gentleness. The crayon marks are still clearly delineated, but their brightness has been scaled down; the skin tones, rendered beautifully in the negative use of the wove Oriental sheet, have a soft, luminous appeal.

31 George Bellows, *The Black Hat* (on white paper), 1921.

The white of the paper may play a more specific role in a print to show a white object. For example, Hilda Volkin uses uninked white paper in the screenprint *Figure I* for the woman's skin and for the patterning on the couch and blanket (see [19]).

Paper may be colored through a variety of techniques. Prepared paper is made by simply applying color to the paper's surface. Colored paper may also be made from colored or dyed rags. Finally, dye or pigment may be added to the vat during the papermaking process. The image and the color often are harmonious. While the image may appear flatter on colored paper, as the crispness of black on white is lost, colored paper has its own attractive, sensuous appeal.

Prepared paper is a term used to identify paper that is colored by coating its surface. The use of prepared paper for drawing occurred as soon as paper was available. In Lorenzo di Credi's *Madonna and Child* [32], ca. 1510, the small sheet, barely five-and-one-half by three-and-one-half inches, and its salmon color are ideally suited to the subject — the tenderness of the mother for her child. The color reinforces the corporeality of both the myriad folds in Mary's garments and the pudginess of the Christ Child. The pointed silver rod used to draw the image embedded minute particles of silver in the paper's surface, which tarnished to delicate, silvery lines. The prepared surface created an ideal ground to receive the silverpoint.

The first step in making prepared paper is preparation of a ground, which facilitates markings from the metal point, lead, or plate. The ground also evens out the surface of the support, be it paper, parchment, or boxwood, and lessens the chance for the stroke of the stylus or raised area of the plate to be interrupted by texture or grain. The ground is prepared by grinding the dry ingredients — white lead, bone dust, seashells, or uncal-

32 Lorenzo di Credi, *Madonna and Child,* ca. 1510.

cinated eggshells — to a fine dust or powder, which is then burnished. Bone dust imparted a gray tone to the coating, eggshells created a warmer cast. Small portions of colored dry pigment, such as salmon or green colors, might be mixed with the basic white of the principal dry ingredients. These dry ingredients were then mixed with a binding medium — gum water, glue water, linseed oil, or saliva — and applied to the support. Today, prepared paper is made from rag paper coated with zinc white suspended in a watercolor or casein binder.[12]

Prepared paper was often used for chiaroscuro drawings, which render forms through dark and light areas and achieve an illusion of depth. In these drawings, which were in vogue from the late fifteenth century through the seventeenth century, the color of the paper is the middle tone. The darker areas of the image were done in metalpoint or ink, while highlights were executed in touches of white or gold.

The Italian artist Ludovico Cardi (1559-1613) used brush and brown ink heightened in white on paper prepared with a green wash for his *Standing Male* [33]. The tonal qualities of light and shadow mold and shape a curved form on a flat surface. The green paper not only softly colors the drawing but also provides a surface that allows the brown ink to recede and the white highlights to stand out, emphasizing depth. The color has started to lift off along the right edge, revealing the paper underneath.

A woodcut technique was later developed to duplicate chiaroscuro drawings on prepared paper. In its simplest form, two blocks were used to print the image. The key block printed the outline of the image in a darker color. A second block printed the color in a middle tone. The blank areas of the paper then became the highlights.

33 Ludovico Cardi, *Standing Male,* ca. 1593.

Colored paper, dating from before the eighteenth century, was made of dyed rags. Despite popular stories of a bag of laundry blueing falling into the vat, these sheets are invariably blue due to the process of manufacture. Indigo was plentiful in Europe, and, like the other natural dyes in use — cochineal and logwood — it was used to dye cloth. It is the only dye that could retain its color throughout the preparatory papermaking processes: the fermentation with lye, the beating, and washing. Even after the cloth was reduced to pulp the blue color was preserved in the fibers used to form a sheet.

Paper made from colored rags has a mottled appearance. Upon close examination of a sheet, individual colored threads can often be distinguished, as in the drawing of *The Fox and the Turkey Hens* [34], dated 1733, by the French artist Jean-Baptiste Oudry (1689-1755). The use of blue paper enhances the artist's theme; heightened by the darker blue watercolor border, it serves effectively as the color of night. The fox, two victims at its feet, stands in a farmyard flooded with moonlight and surveys three other hens safely perched in the tree. The illusion of depth is successfully rendered by the black ink on a middle-toned sheet. White highlights in the moon, the fox's fur, the tree, and the farmhouse complete the realistically portrayed scene.

Colors other than blue became more common in handmade paper after the eighteenth century because of two factors. The Hollander beater eliminated the fermentation process that destroyed colors other than indigo. Chemistry, which developed as a science in the late eighteenth century, created other dyes for cloth. These colored cloths when used as raw material for pulp survived the maceration process.

Artists such as Camille Pissarro (1830-1903) took advantage of the new range of colors presented by these papers. Pissarro carefully chose his papers to enhance

34 Jean-Baptiste Oudry, *The Fox and the Turkey Hens*, 1733.

35 Camille Pissarro, *Nude Women,* 1922.

36 Pissarro, *Nude Women,* 1923.

his imagery. His series of works illustrating forest scenes, for example, was printed on green paper. The effect of differing colors may be seen in his two lithographs of *Nude Women* [35 & 36]. One is printed on smooth, wove white paper [35] and exhibits a crisp contrast between black ink and white sheet. Although the other [36] was printed posthumously on pink paper, the choice of the colored sheet was in keeping with Pissarro's aesthetic sense. The pink sheet is ideal for the subject of nude bathers: it colors the skin of the three women and creates an overall sensuous, voluptuous effect. The red and blue colors of the rags used to make the pulp are still visible as individual threads in the sheet. The individual elements of color harmonize when the print is viewed from a distance and recalls the blending of colors in Pissarro's Impressionist painting.[13]

The third way of coloring paper is by adding dye or pigment to the pulp. This technique, practiced since the end of the eighteenth century, is now the most common method used by commercial mills to color paper. Chemistry again played a role in this development by producing new dyes and pigments. It was, however, the spread of the Hollander beater during the eighteenth century that facilitated this method. The beater allowed for continuous, even agitation and mixing of the added color. Since the mechanism was already in place, the practice of mixing color in the beater spread quickly after the technique had been tried. Unlike the mottled appearance of paper made from dyed rags, the tint of paper manufactured with this method is even throughout the sheet.

Pigments are convenient and easy to use. The beater is loaded with water and rags. Dry pigment, mixed with hot water, is then added to the beater, which mixes the pigment with the rags. After the color is mixed, a retention agent is then added to the pulp; this enables the pigment molecules to find their place in the interstices of the fiber.

Pigments may be used to make sheets that complement the subject matter. In her book of poems, *Making a Sacher Torte* (published in 1981 by the Perishable Press and illustrated by Ellen Lanyon), the contemporary writer Diane Wakoski focuses on food — a green tomato pie, olives, mushrooms, milk, or plums [37]. The Hamadys made colored paper and selected ink to correspond to the imagery of the poems. The illustration accompanying "Sally Plum," for example, appears in purple ink on a pink sheet made from dyed rags. The poem itself is printed in gray ink on purple paper colored with pigment [38]. The difference in the two sheets is immediately evident. The pink sheet has a mottled appearance that lends variety to the surface of the paper and reveals its origins: red and blue rags. The purple, an intense color distributed evenly throughout the sheet, is an elegant and appropriate choice for the "plum" poem.

Size

The actual size of a paper sheet is less significant than the width of the margins and its relationship to the image. Most Old Master prints did not have large margins. Collectors purposely trimmed the margin to the plate mark so a print could be displayed in an album. The margin might also be trimmed if it had become dirty through handling.

Early bookmakers, however, who adopted the practice of decorated margins from illuminated manuscripts, viewed the blank area around the text as an essential element of page design. As can be seen in the seventeenth-century *Florilegium* [3], the borders may be purely decorative or may continue the central theme (in this case, floral) to the edges of the page. The border is

37 Walter Hamady, frontispiece for *Making a Sacher Torte,* 1981.

38 Hamady, page from *Making a Sacher Torte.*

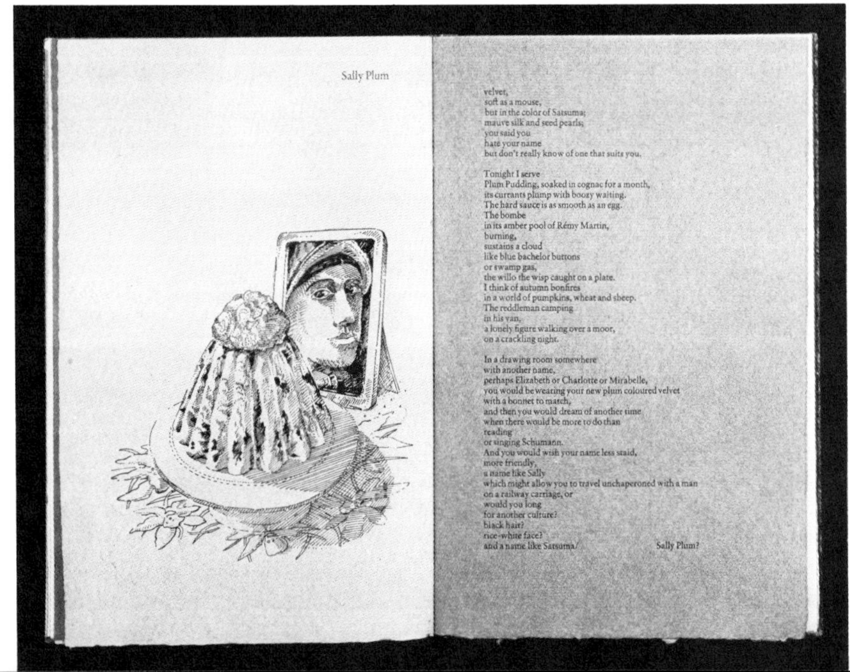

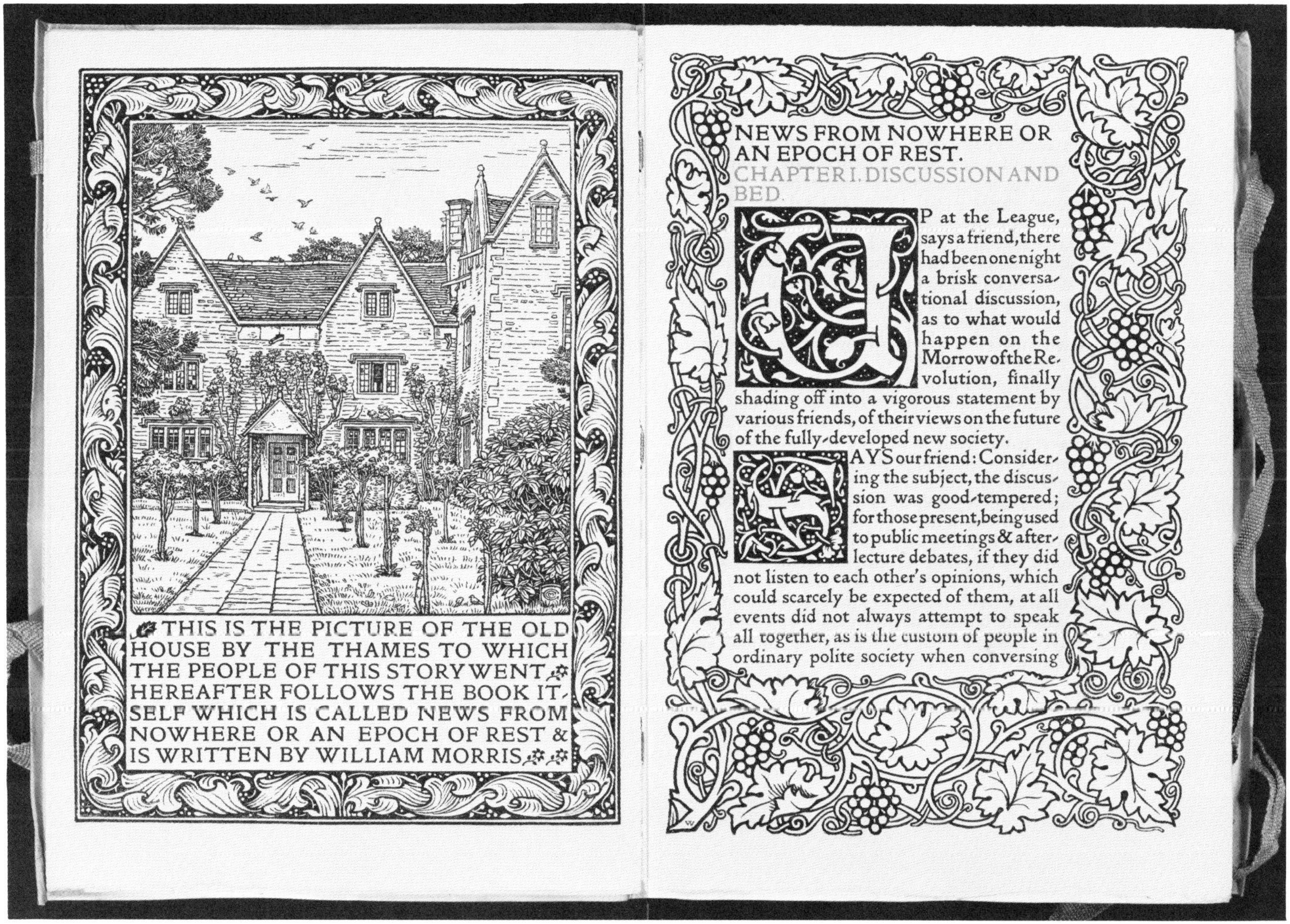

THIS IS THE PICTURE OF THE OLD HOUSE BY THE THAMES TO WHICH THE PEOPLE OF THIS STORY WENT. HEREAFTER FOLLOWS THE BOOK ITSELF WHICH IS CALLED NEWS FROM NOWHERE OR AN EPOCH OF REST & IS WRITTEN BY WILLIAM MORRIS.

NEWS FROM NOWHERE OR AN EPOCH OF REST.

CHAPTER I. DISCUSSION AND BED.

UP at the League, says a friend, there had been one night a brisk conversational discussion, as to what would happen on the Morrow of the Revolution, finally shading off into a vigorous statement by various friends, of their views on the future of the fully-developed new society.

SAYS our friend: Considering the subject, the discussion was good-tempered; for those present, being used to public meetings & after-lecture debates, if they did not listen to each other's opinions, which could scarcely be expected of them, at all events did not always attempt to speak all together, as is the custom of people in ordinary polite society when conversing

39 William Morris, title page (left) and facing page from *News from Nowhere or an Epoch of Rest,* 1892.

clearly segregated from the image by a straight outline of black ink; its width increases in size as it encircles the central image.

William Morris revived the medieval and Renaissance treatment of letterforms and borders in his handmade books produced in the late nineteenth century. His utopian novel *News from Nowhere or an Epoch of Rest,* published by him in 1892, is characteristic of the Kelmscott book designs. Placement of type and illustration on the page is of primary importance; the two facing pages are regarded as one unit. The size of the margin increases from the gutter (where the book is bound) to the top, side, and bottom of the page. A Gothic-style house surrounded by trees and shrubs adorns the title page [39]. The illustration and text, set in a block, are enclosed by a border design of stylized leaves set against a black ground. The first lines of Chapter One on the facing page are printed in red ink. The beginning letter of each paragraph is greatly enlarged and highly decorative. The letters "U" and "S" are encircled by tendrils and accented with white dots and flower buds. The type on this page is also set in a block, but the border is airier. Even though the borders of both pages vary considerably, their decorative designs complement the text and illustrations. Morris's unification of borders, typeface, and illustrations derives ultimately from centuries-old sources.[14]

In the early nineteenth century the margins of the sheet became important to collectors in two ways. First, the deckle edge in books became fashionable. Bookmakers had traditionally viewed the deckle edge as a disadvantage; printers cut it off to give an exact edge for registration in the press; and if the ragged edge escaped the printer, the binder was apt to trim it. Nineteenth-century collectors, however, came to view the deckle edge with reverence, since it was for them a reminder of the process of forming a sheet. Later in the same century a machine method was invented to reproduce this result of hand papermaking.

Secondly, there developed a fervor for larger margins in Old Master prints. The collector Joseph Maberly advised against trimming a print and argued for leaving the "margin sanctuary" entirely alone.[15] Margins were even added to prints by mounting them on a larger sheet of paper. New plate marks were sometimes printed on these mounted sheets.

The positive treatment of margins began in the late nineteenth century. The use of figurative borders in single prints occurs most frequently in the "remarques" (or sketches) made along the edge of a printing plate to test the acid bath or lithograph crayon, or to "set the hand" before working on the central image. Usually removed, they can indicate early proofs of a print or serve as a way to see an artist's notation on the progress of the plate. During the mid-nineteenth century, these remarques eventually came to be viewed as separate decorations that either altered or enhanced the blank margin and were allowed to remain on the plate and therefore to be printed with the central image. Sometimes they were created intentionally in drawings of small details that related to the central image or to distinguish desirable early proofs. The French artist Félix Hilaire Buhot (1847-1898), who frequently created small drawings on the edge of the sheet, called them *remarques symphoniques.*[16]

The third and fifth states of Buhot's etching *Country Neighbors* [40 & 41] demonstrate the changes he made as the printing progressed. (Each change in the plate indicates a separate state.) In each print two country folk make their way through a storm. The wind whips their wraps around their bodies and bends their sheltering umbrella. Light shimmers on the puddle at their feet,

40 Félix Hilaire Buhot, *The Country Neighbors,* state III.

41 Buhot, *The Country Neighbors,* state V.

the fence posts, and along the edges of the road. In the third state [40] the area around the central image contains only a few remarques. A woman's face, two peasants, and a curled-up dog are visible at the left margin. In the fifth state [41] these are drawn in greater detail. In particular, the woman's face now resembles a miniature portrait in an elaborate frame. This design is in sharp contrast to the two country folk of the central image who face inclement weather in their heavy wraps and wooden clogs. The portrait is more in keeping with the aristocratic, frivolous rococo imagery of the eighteenth century than this realist genre scene popular 100 years later. Further, in the fifth state a shell with garlands, another remarque recalling rococo motifs, is added; and the right border is filled with sketch marks resembling small faces and a ram's head. Although some of the remarques relate to the image, such as the sketch of a peasant couple, others do not. A comparison of these two states illustrates the way in which the border affects how an image is perceived.

James McNeill Whistler (1834-1903) was the first artist to make an aesthetic issue of the size of print margins. He abhorred the practice of adding remarques to the final etched image and usually clipped the print right to the plate mark — as in *The Garden* [42]. He left a small tab on which to draw his butterfly insignia. In an attempt to alleviate any competition with the image, these tabs gradually became smaller in size. The preciousness and self-consciousness of the "butterfly" tab seem to contradict his purpose. The more collectors craved margins, the more Whistler clipped them. As collectors became more and more obsessed with signatures, Whistler's became smaller.

The positive use of margins so that the image "floats" on the sheet may, however, have begun with Whistler's lithographs of the 1880s. His sparse, sketchy images of trees, figures, and buildings were allowed to float on a large expanse of white paper that served as foreground, horizon, clouds, and sky.

Jasper Johns (b. 1930) continues this positive use of the white paper in his work. In *Numerals* [43], done in 1975, two rows of numbers float on a large sheet of untrimmed paper. Each number — its style and size reminiscent of children's blocks — was printed from a separate copper plate; each was aligned very closely to the next so that the imprint of all ten plates formed a gridlike pattern. The white of the paper that is visible in each number creates a visual tie between the ten blocks as well as to the large expanse of paper. Part of the beauty of the print is the tonal range and variety of ways in which the numbers are delineated. The precise treatment of the overall print contrasts with the everyday subject matter of the image. The fine, large white sheet with its deckle edge, the image carefully proportioned to the space surrounding it, and the skillful printing create an exquisite piece.

42 James McNeill Whistler, *The Garden*, 1886.

43 Jasper Johns, *Numerals 0-9*, 1975.

1. This classification of the characteristics of paper derives from Andrew Robison's exhibition catalog, *Paper in Prints* (Washington, D.C.: National Gallery of Art, 1977). I am indebted to him not only for his method of organization but also in what to look for within the given categories.
2. Dard Hunter, *Papermaking, The History & Technique of an Ancient Craft* (New York: Dover Publications, 1978), p. 16. Also see Edward Lucie-Smith, *The Story of Craft, The Craftsman's Role in Society* (Ithaca, New York: Cornell/Phaidon Books, 1981), p. 139: "A folio volume consisting of 200 pages of parchment needed the hides of no fewer than twenty-five sheep — the scribe's labour in producing a handwritten manuscript has always been the smaller part of its cost."
3. Francis Ames-Lewis, *Drawing in Early Renaissance Italy* (New Haven & London: Yale University Press, 1981), p. 22.
4. Leonard Schlosser, "The Graphic Confluence of 1800," in *A Miscellany for Bibliophiles*, ed. H. George Fletcher (New York: Grastorf & Lang, 1981), p. 84.
5. Schlosser, "A History of Paper," in *Paper – Art & Technology* (San Francisco: World Print Council, 1979), p. 17.
6. Ibid., p. 18.
7. For a complete description of *cliché-verre* see Osbert H. Barnard, "The Clichés-verres of the Barbizon School," *The Print Collectors Quarterly*, Vol. 9 (1922), 149-72.
8. The stock used by the Kodak Company in making its paper comes from Canada in the form of wood pulp. The pulp is put in a hydromixer, rewashed, and then cast on very fine mesh screens. The speed that the screen goes under the hopper full of pulp is what determines the thickness of the sheet — categorized as double, medium, or single weight.

A coating of baryta, a product of barium and similar to sizing, fills in the "hills" and "valleys" of the sheet, creating a smooth surface. The coating will warm the paper's tone if a creamy tint is added to the baryta. In this type of paper the highlights in the printed image exhibit an additional warmth. Kodak's Ektalure, often used for portraits, is one such warm-toned paper. RC or "resin-coated" paper is coated on both sides with synthetic resin. The resin does not absorb the photochemicals and drastically reduces processing time.

Smooth or glossy surfaced paper is highly calendered. If the paper is to have a heavily textured surface, it is fed through textured calender rollers, which reintroduce texture to the sheet. This type of paper may be called "tweed" or "tapestry."

A light-sensitive emulsion is then flowcoated over the sheet. The color of the emulsion's silver deposits determines the image tone of a print. It may be warm (brown-black) or cold (blue-black). Agfa's Portriga-Rapid is one type of paper with a distinct brown tone. The emulsion tone and paper tone are harmonized; a warm-toned emulsion coats a creamy sheet.

I am grateful to Arthur Barnes, product specialist at the Eastman Kodak Company, Rochester, New York, for this information.

9. Made in 1964, this photograph, printed on smooth-surfaced glossy paper, embodies the standard set by Ansel Adams and Edward Weston. Uelsmann's choice of this paper reflects the concern for the tonal range and brilliance that such a paper offers. Uelsmann believes that this smooth surface worked against photography's acceptance as a fine art in the 1950s, since the Abstract Expressionist aesthetic (dominant at that time) of creating a thick surface of paint that records the artist's gestures was not possible in photography. (From a telephone conversation with the artist in June, 1982).

This may also be one reason for the revival in the 1970s of the early twentieth-century techniques that introduced color and surface texture into the photograph, such as the gum-bichromate printing method.

10. *Print Collector's Newsletter*, Vol. 10, No. 3 (July-August 1979), 94.

11. Robison, *Paper in Prints*, p. 23.

12. James Watrous, *The Craft of Old-Master Drawings* (Madison: University of Wisconsin Press, 1957), pp. 13-15.

13. These two lithographs are excellent examples of the difficulty presented by isolating the individual elements that influence the final image — surface, color, and size. Texture is a critical factor in each print. The smooth, wove white sheet [35] separates and defines the darker areas, as is evident in the middle figure's head and hair. The wove surface holds the crayon marks, so the eye easily distinguishes the head from the background foliage. It also provides a smooth finish for the women's bodies. In contrast, the pink paper [36] has a laid surface, which interrupts and also absorbs these fine details. Here, the darker areas fuse together and are more difficult to read.

Finally, the same plate was printed on two different sizes of paper. The image floats on the white sheet, giving breathing space to the figures. The image on the pink paper, however, has been cropped right to the edge — crowding the figures. Further, the placement and therefore the symmetry of the image on the pink paper is different from the other print.

14. For a complete description of the elements of Morris's designs see Susan Otis Thompson, *American Book Design and William Morris* (New York & London: R. R. Bowker, 1977).

15. Joseph Maberly, *The Print Collector*, 2d ed. (New York, 1880), p. 43.

16. Robison, *Paper in Prints*, pp. 44-45.

Paper as a Contemporary Art Form

Contemporary artists use paper not only as a support for prints or drawings but also as an inherently expressive art material with possibilities as limitless as their own imaginations. They see paper as an active participant that may be molded, sculpted, collaged, scarred, or wrapped.

The willingness to look at the potential of three-dimensionality in paper is not new to either the East or West. In fifteenth-century Florence, designs stamped on metal armor were first proofed on paper molded to the shape of the armor before being etched. In eighteenth-century Japan, artists embossed their woodcuts to create subtly raised designs (*gauffrage*). The French sculptors Aristide Rousseau, Maurice Dumond, and Pierre Roche, working during the late nineteenth and early twentieth centuries, used plaster and steel molds to emboss a normally flat piece of paper.[1]

In general, however, the traditional Western attitude toward paper as merely a support for an artistic image remained unaltered until the twentieth century. Paper was viewed as a transient material, a cheap malleable substance used by the artist for initial sketches and studies. Paper came to perform a structural and conceptual role in *papiers collés*, first as employed by Georges Braque and Pablo Picasso in their Cubist works and later by the Russian constructivists. A scrap of newsprint cut in the shape of a violin and pasted into the composition of a drawing or painting retains some of its original appearance and function. It refers to a former purpose outside the confines of the frame: transmission of information, perhaps banal, now dated. The Cubists, however, were not concerned with elevating paper as an artistic material; they used this inexpensive and available substance in a purely formal way. Later, the Surrealists more deliberately combined cut-up images for their references to worldly matters. Kurt Schwitters, who created his own form of Dada which he termed "Merz," adapted the Cubist technique of collage to his own purposes. His series of "Anna Blume" poems was made up of selected sentimental clichés from scraps of printed materials.

In the 1920s at the Bauhaus, an art school in Germany, Joseph Albers included in the *Vorkurs* (beginning classes) an investigation of paper and its properties. Students were given a sheet of paper that they might cut, weave, sculpt, or glue in order to discover the artistic potential of paper. Their pieces were not meant to be art works but rather to teach students to work with paper's fragility, strength, and malleability. The results of these experiments, which were published along with the other descriptions of courses at the Bauhaus, enlightened artists to the creative potential of paper.

The metal collage prints of Rolf Nesch produced during the 1920s encouraged a sculptural approach to printmaking. Later, in the 1940s, the work of the famous printmaker Stanley William Hayter introduced level in a traditionally flat medium; his work still influences contemporary printmakers. Hayter experimented with

44 Inez Storer, *Wrapped I*, 1976.

deeply incised lines on a lithograph stone, which when printed caused the ink to literally stand up on the sheet so the white paper areas appear to be in "a valley" or trough.[2]

In the 1940s and 1950s paper became more fully appreciated when Abstract Expressionists such as Franz Kline and Willem de Kooning, who, in considering paper a suitable support for their work, began to use it like canvas. Jackson Pollock, working with Douglass Howell in the 1950s, used handmade paper for drawings with ink and water colors that blend right into the paper.[3] Of the Abstract Expressionists, Robert Motherwell probably had the most direct contact with the Surrealists. In his work, he synthesized both the Cubist and Surrealist approaches to collage. Like Braque and Picasso, Motherwell used paper to create abstract images with visual references to outside meanings, as allusions to the unconscious experience. The "fortuitous association on a single plane of disparate elements" could provide a level of generative force equal to psychic automatism.[4]

In 1974 Robert Rauschenberg traveled to the Richard du Bas Mill in Ambert, France, to create his series *Pages and Fuses*. These handmade paper multiples combine printmaking and sculpture, and they further expanded the paper vocabulary. Centers of papermaking had already been established in the United States by the time Rauschenberg went to France. These centers met the needs of both students and professionals in making paper or in combining it with other media.

Contemporary paper pieces span the spectrum of modern art. Artists who work in paper frequently come from other media, such as sculpture, painting, textiles, and graphics. They may transpose the forms and images of these other media into paper, or they may work only with paper exploring its rich tradition and infinite capabilities. *Wrapped I* [44] by Inez Storer (b. 1933) demonstrates the multiple ways in which paper is generally used — in books, lists, and newspapers. Embedded in the center of this work is a collage made up of a faded mauve book cover, a scrap of newspaper, and a list of household objects: "sheets, spreads, glass cloths, tea cloths . . . ." A vertically hung piece of paper visually affixes the collage package to the background sheet, while golden threads further the wrapping. The threads trail off the bottom of the collage, their random fall echoed by drawn pencil lines. Color, added in and around the collage, appears to yellow and age the piece.

The barely visible old style lettering and a list of common articles that somehow relate to domestic chores and pleasures — books, ironing — heighten the nostalgic appeal of *Wrapped I*. Information is partially obscured and buried, yet partially revealed, as if from a tattered diary found in an attic. Two rectangular embossed lines above and below the collage preciously frame and complete *Wrapped I*.

William Weege (b. 1935) in *Oconomowoc* [45] delights in the tactile qualities of pulp and the sensual appeal of the fibers. Named after a town in Wisconsin, *Oconomowoc* was made at both the Jones Road Print Shop and the Upper United States Paper Mill. It separates yet unifies the basic elements used in forming a sheet — thread and pulp. The work consists of a grid of colored string that was laid and pressed onto a white sheet of paper. In fabricating the work, Weege strung twine horizontally across a wooden frame in an even pattern. Paper pulp clung to these strings when the wooden frame was dipped in a vat. The horizontal strings were then painted with a light blue acrylic. The frame was then placed on the white sheet and the ends of the string cut, releasing the colored strands onto the white paper sheet.

The more brilliantly colored vertical strands of string and pulp were similarly formed and painted and then placed on top of the horizontal grid. After the whole sheet was sprinkled with glitter, it was covered with a felt, then pressed and dried. The web of strings now embedded in the sheet forms a grid of colorful pulp that is, and yet is not, part of the sheet.

45 William Weege III, *Oconomowoc*, 1979.

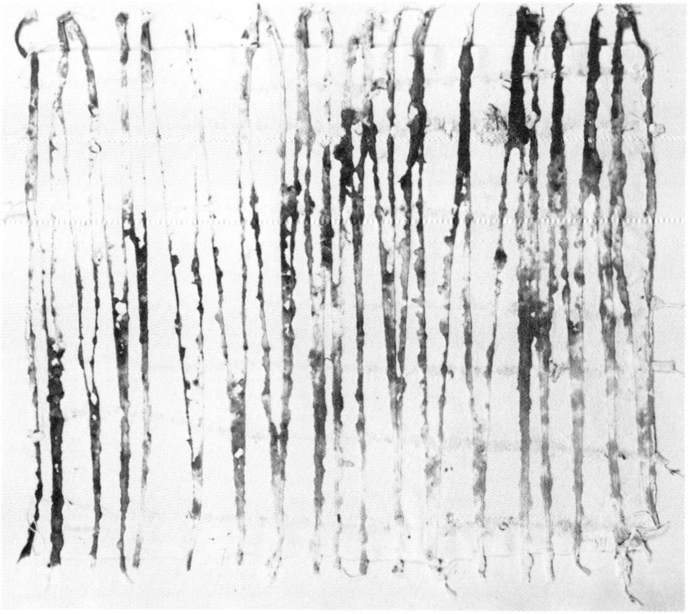

46 Suzanne Anker, *Ribbed Series, #3*, 1975.

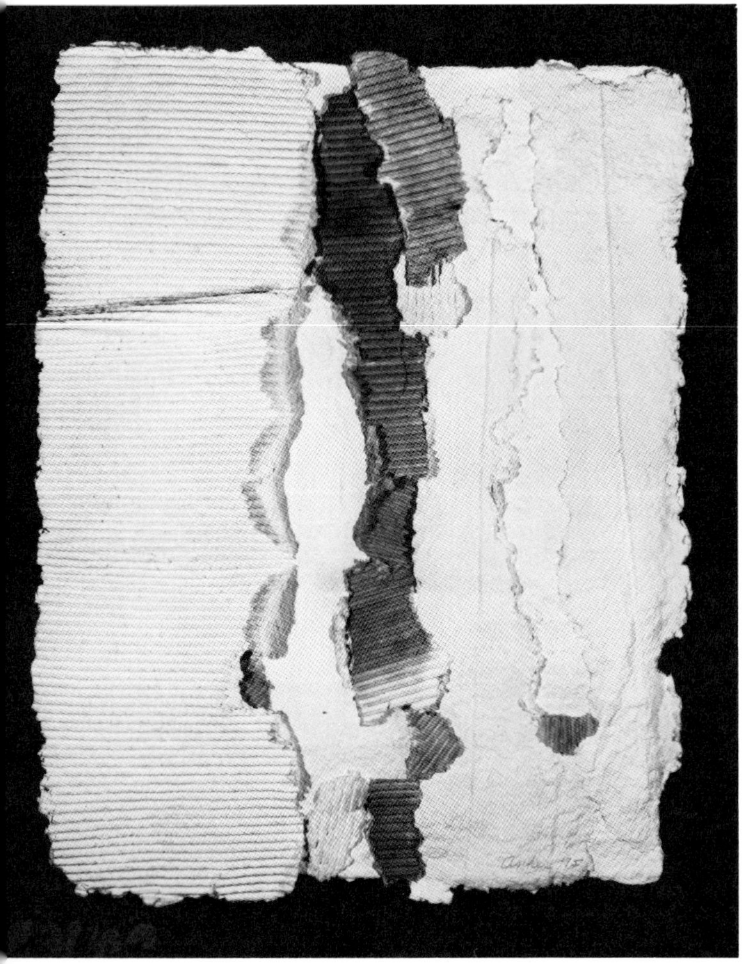

Why would an artist choose to work in paper? The contemporary papermaker Garner Tullis says in reference to the nineteenth-century French sculptors who, accustomed to three-dimensional forms, embossed prints from molds: "It wasn't that paper was something new and exciting. It was just that it served a specific need, an intellectual demand that would enhance what they wanted to say in their work."[5]

The willingness to extend the limits of any media is encouraged in modern art. The painter Kenneth Noland remarks about the gradual elimination of drawing in the creation of images in the 1950s and 1960s: "As that has been withdrawn by some artists, a different kind of emphasis has taken place . . . [having] more to do with tactile relations in handling stuff. I am referring to the interest in collages as an example — taking things and pasting and sticking them instead of having to depict things by drawing. This new interest gives artists access to the stuff."[6]

The properties of the "stuff" (pulp) have a strong attraction. Noland describes the pulp as "something I could handle in the same way as wet paint and raw cotton duck."[7] For the sculptor Suzanne Anker pulp is a "receptive material similar to clay" that allows her to deal with sculptural concerns of form and light while her hand is allowed "to run free with applied paint and pastel."[8]

Anker (b. 1946) used corrugated cardboard in the construction of *Ribbed Series, #3* [46]. Pulp was pressed by hand into a latex rubber mold taken from a sheet of cardboard and the water was gradually extracted with a sponge. These pieces are then torn, shaped, colored, and collaged onto a warped sheet of handmade white paper, creating a paper relief. An unribbed creamy sheet was then laminated on the right half. The colors in the ribbed forms modulate from a deep salmon to a light pink; where the color is darkest, it has bled onto the paper underneath. *Ribbed Series, #3* stresses the tension between strong verticals in the torn forms contrasted by the horizontal ribbings. A strong band of color, composed of small ribbed pieces, dissects her work. The piece is a subtle play of textures, shape, compositional movement, and color values that recall the landscape of California where Anker lives.

Richard Hunt (b. 1931), as a sculptor, brings his concern for dealing with materials by allowing them to reveal their unique properties to paper. *Paper Piece II* [47] emphasizes the simple fact that two sheets laid together without felts will bond. A gray sheet forms the bottom layer, to which unevenly mixed white pulp gives a mottled effect. Smooth paper couched on top forms abstract geometric shapes that float like islands and selectively reveal the mottled pulp underneath.

Perhaps the most complete and exhaustive list of the ways an artist can manipulate paper has been set forth by Jules Heller: "Paper — when employed by artists — may be torn, cut, folded, spindled, or mutilated in many ways. It may be burned, scorched, exploded, waxed, dyed, glued, curled, sewn, taped, shaped, stamped, heated, cast, rolled, distressed, crumpled, sawn (as wood), drawn upon, or painted. It may be used as supports for lithographs, intaglios, relief or screen prints. It may be used for collage, assemblage, frottage, *papiers collés*, or three dimensionally as pulp in molds. It may be recycled again and again, or used as a receptacle, or for embedments. It may be burnished, hammered, or embellished with fibers, feathers, golf and silver leaf, wire, or found materials. It may be made waterproof, fireproof, insect-proof, couched (sheet upon sheet), and vacuum-formed."[9]

When artists discuss pulp, their language often takes a reverential bent, for there is both a spiritual as well as

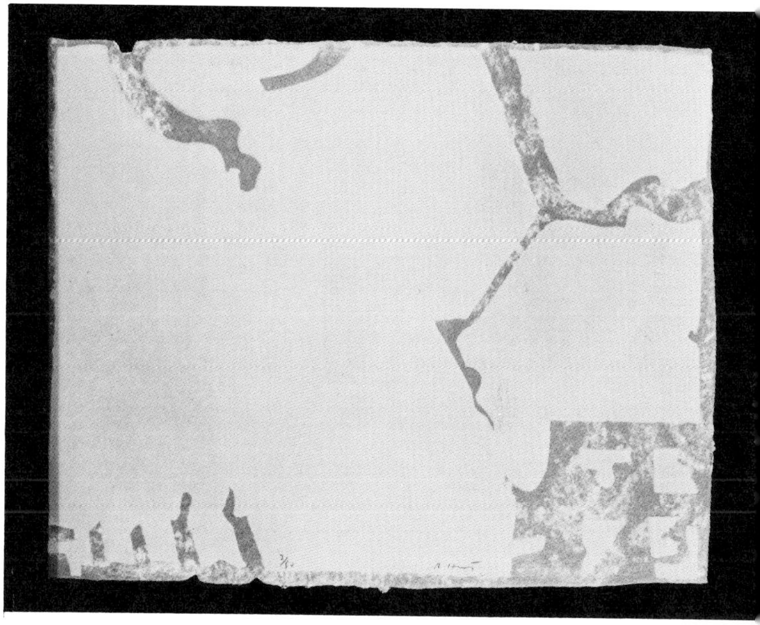

47 Richard Hunt, *Paper Piece II (Untitled)*, 1978.

physical attraction to paper. Kenneth Noland became interested in paper only when he "got his hands wet"; he discovered "something that hovers around the threshold of the surface" that the papermaker can get underneath: "You can get in it and make something come up and become the surface."[10] The nature of the fibers in water, tactile and sensuous, catches the imagination and leads to the choice of pulp as the artist's material. "In the first encounter, time must be allowed to get past the inevitable attraction of the medium's force."[11]

Harold Paris (1925-1979), a sculptor from the Bay area of California, is an artist who has "gotten beneath the

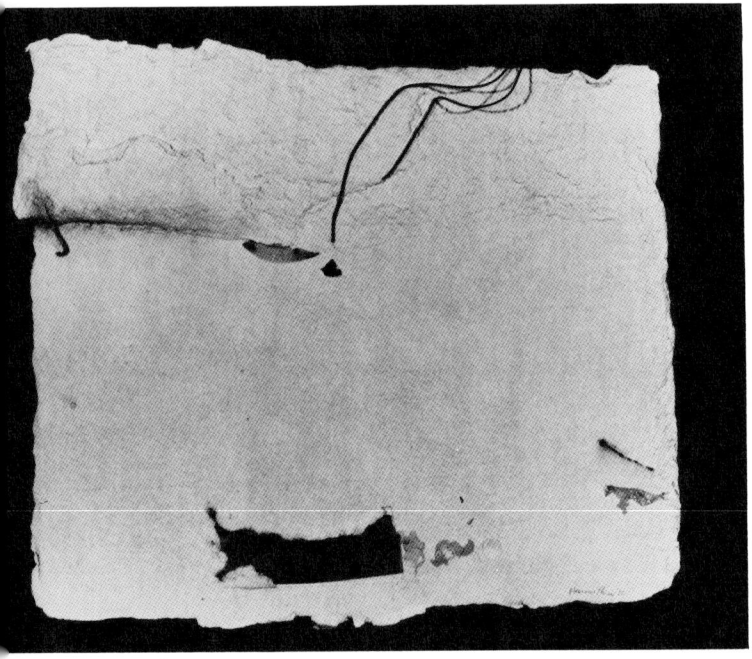

48 Harold Paris, *Carta Series, #50,* 1978.

surface." During the mid-1970s he produced a set of fifty paper pieces called the "Carta Series." In executing these, Paris took advantage of the facilities at the Institute for Experimental Printmaking in Santa Cruz and the assistance of its director, Garner Tullis. Paris had been creating "magical little tableaux by embedding mysterious objects in small panels of laminated silicone gel flooded with translucent colors."[12] The "Cartas" produced at Santa Cruz were a logical next step. The paper pulp was impregnated with autobiographical mementos — bits of confetti, newspaper, wire, cellophane — and run through a hydraulic press, sometimes up to six times, to build up the dense layers. Paris called the process of making the pulp, casting the sheet, embedding the images, and coloring, a "ritual which not only immersed the object but portions of myself."[13]

The top of *Carta Series, #50* [48] contains an added section of paper that is folded back. A piece of wire embedded inside "bled" out through the surface of the paper, showing the wire's hidden presence. The upper portion of the work may have been sponged to create extra texture, while small bits of confetti and cellophane accent the lower portion. A piece of brown paper was positioned on the lower edge, then shifted slightly. The pulp remembers the first placement and retains its linear outline. This use of paper in a personal, intimate, and nostalgic way is also evident in the work of Inez Storer (see [44]), a former student of Paris.

June Bonner (b. 1918), a Cleveland artist, in *Untitled Paper 2* [49], also utilizes that space that hovers below the surface. Cotton ropes, embedded between two sheets of paper, undulate from left to right and bits of leaves, thrown into the vat, remain as specks of plant material in the linen sheet. The bottom corners were dipped three times in indigo dye in a motion that continues the rhythm of the cotton ropes. The overlapping waves of color mysteriously create white lines that reiterate the rhythm of color and string.

The current use of this "fascinating material" was aided by two men who preserved the knowledge and skills of hand papermaking: Dard Hunter and Douglass Howell. In numerous books and articles, Dard Hunter outlined the history, methods, and materials of the ancient craft of papermaking. At one time his was the only commercial hand paper mill in the United States. It closed in

1931, after only three years of operation. In 1939 he established the Dard Hunter Paper Museum, now located in Appleton, Wisconsin. It houses his collection, amassed from a lifetime of interest in paper. Hunter's scholarship, enthusiasm, and dedication have inspired and informed the succeeding generations of papermakers.

If Hunter is credited with the renaissance of traditional papermaking, Douglass Howell is considered the spiritual leader of papermaking. Now seventy-six years old, Howell has kept hand papermaking alive from the 1950s. As a youth his love of papermaking prospered under the tutelage of Guido Biaggi, director of the Biblioteca Lorenziana in Florence, Italy, where he spent several years. While in Florence, he also had the opportunity to see the paper used by the Italian Renaissance masters for their drawings in the Uffizi Gallery. His innovative spirit is evident in the special papers he has manufactured for such artists as Ann Ryan and Jackson Pollock. Always curious, he analyzes different fibers and experiments with them in his laboratory and workshop on Long Island. He uses his own paper pulp to construct his own "papetries."[14] He also has trained papermakers in his studio — among them, Laurence Barker (b. 1930).

In 1964 Barker set up facilities at Cranbrook, Michigan. Originally intending to make handmade paper of high quality for use in printmaking, Barker himself began integrating his intaglios into a more sculptural sheet. *Untitled #5* [50], a fine example of the cross-fertilization with other arts, consists of a large gray sheet that has been distressed and folded. Embedded string emphasizes the rhythm of the creased areas. One piece of string, looping around itself, almost dissects the piece diagonally. Pencil and crayon marks, added along with ink and a wash of darker blue pulp, echo the foldings and the embedded string. A rectangular piece of paper made of the same pulp was embossed, inked, and sewn onto the surface. A hole at the lower right shows the imprint of the artist's finger that lifted away the pulp. A horizontal blue sheet laminated underneath the gray picks up the diagonal line of blue pulp and an orange crayon squiggle. Graphite drawn over a stencil of "Yds" balances the inked rectangle and refers perhaps to yards of fabric — one of the raw materials used to make this and other paper pieces.

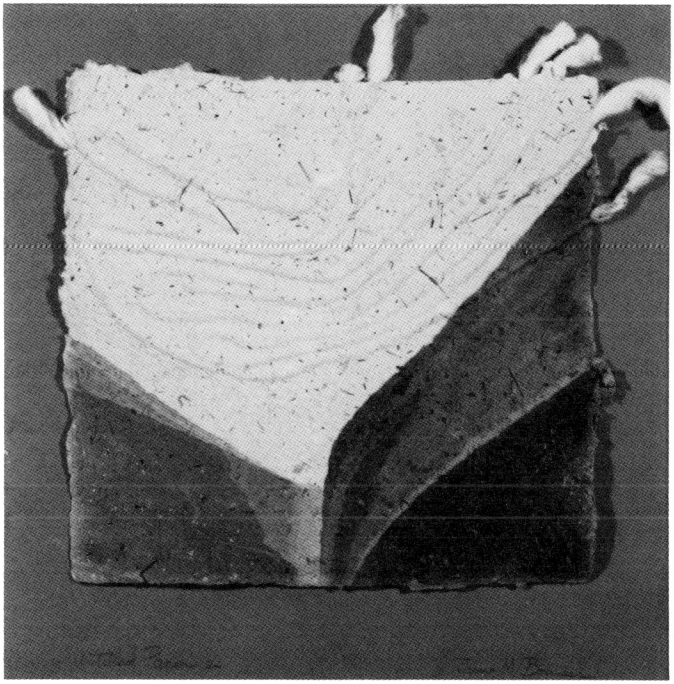

49 June Bonner, *Untitled Paper 2*, 1979.

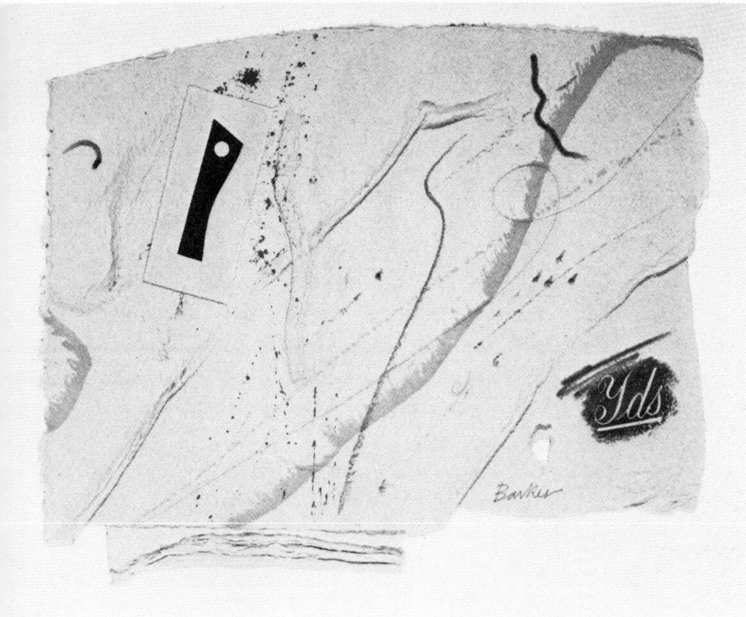

50 Laurence Barker, *Untitled #5.* See also color plate.

Barker's view is that it is "precisely in the hyphen between support and medium, where ground becomes figure that much untapped poetry resides, acting paper that is in turn acted upon. Paper projects itself in shape, texture, and color as a medium and simultaneously is a vehicle for drawing, painting, and printing. Like a rheostat, paper can easily be dialed down, as it were, to the near-zero assertiveness of pure support — i.e., sheets that are white, smooth, and square — as it can be cranked up through stages to an intensity of activity that is pure medium."[15]

Between 1964 and 1968 Barker trained a number of people who have played major roles in the revival of hand papermaking. In the Midwest they include Walter Hamady, who set up facilities at the University of Wisconsin; Roland Poska, of the Fishy Whale Press and Mill in Illinois; Avis Koutroulis, at Wayne State University in Detroit; and in the East, Winifred Lutz at Yale and John Koller of H. M. P. (Company) in Connecticut.

Like the ripples on a pond when a rock is thrown into it, the circles of influence spread outward. Kathryn Clark, who studied with Avis Koutroulis, and her husband Howard, in 1972 set up Twinrocker, Inc., in Brookston, Indiana. Joe Wilfer, a former student of Walter Hamady, established the Upper U. S. Paper Mill in 1975 in Oregon, Wisconsin; now disbanded, it served as a center of collaborative activity related to William Weege's Jones Road Print Shop (see [45]).

In 1973 Ann and Garner Tullis started the Institute for Experimental Printmaking, now simply called Experimental Printmaking and relocated in San Francisco. Dedicated to stretching the limits of technology, the Institute strives to meet the artist's ends, to fabricate works that "manifest the hands of the artist instead of the machines, and to do it harmoniously, successfully, and contributively in the latter part of the twentieth century."[16] Available to the artist are letterpress, vacuum, screenprint, etching, and lithograph presses and a paper mill. Artists who took advantage of these facilities include Inez Storer (see [44]), Harold Paris (see [48]), Suzanne Anker (see [46]), and Claire Falkenstein.

Falkenstein created *Never Ending Stream* [51] at the Institute in 1975. Thin, semi-translucent, pure white pulp was cast onto a mold to form a three-dimensional maze. The ragged and fragile walls of the maze, one-half inch in height, were then manipulated by hand. The depressed areas inside the maze were smoothed by hand in contrast to the untouched outer edges. The maze is self-contained; its thin walls stand up crisply against the base sheet. Pure white antler-or-branch-like forms within the circular screen produce a subtle play of height and shadow.

Garner Tullis (b. 1939) finds time not only to advise other artists but to do his own work. In 1974 he took over the workshop to execute his "technical artifacts." For some of these he used the Hollander beater to recycle his student Abstract Expressionist paintings into pulp. The pulp, after being mixed with water to a consistency of thin gruel, was then strained for a thicker consistency. It was patted onto molds (which Tullis had used earlier for dimensional prints) and hand pressed with a sponge for detail. A similar process utilizing rags produced the high-relief print *Technological Artifact – Indian* [52]. Like his other technical artifacts, the piece was covered with graphite. Although the graphite creates a lead-like visage, the ragged edge of the paper subtly reminds the viewer of its fragile, malleable nature. Tullis scarred the pliable pulp with feather-like patterns and long strokes that resemble a quiver of arrows. "I wanted to make a statement about paper — working with the oldest material related to printmaking, but in a new way. And I wanted to defy the use of all equipment that I had in the shop, to do something that was almost entirely handmade. Nathan's [Nathan Oliveira] pieces stimulated me to use the graphite. If he could use lead like paper, why couldn't I make paper look like lead?"[17] *Technological Artifact – Indian* is indeed more than a paper mask: it resembles a fossil in stone, the power and fierceness in the face conveyed as if through time.

A similar sculptural approach used to very different effect is seen in *Jory I* [53]. Here the material is plastic, which has been vacuum-formed over a mold. The sense of foreboding expressed in the smooth white plastic face is increased by the placid demeanor of its features. Tullis's exploration of new and unusual techniques and his desire to make the artist's hand visible is evident in

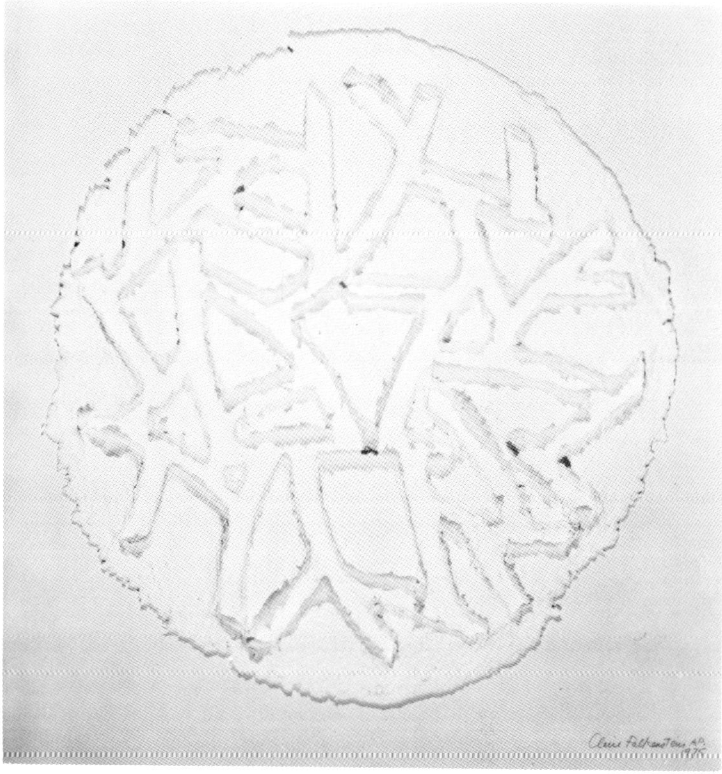

51 Claire Falkenstein, *Never Ending Screen*, 1975.

both these objects. He transforms paper and plastic from a normally inactive, flat surface to an active material that is both solid and dimensional.

As artists and proprietors of "facilitating workshops," people like Garner Tullis are able to tailor their facilities to the needs of the artist. Today, there is a general interest in expanding the traditional print vocabulary and in closing the gap between art and craft. As artists' interest in the "mechanics of process"[18] flourish, the

52 Garner Tullis, *Technological Artifact – Indian,* 1975.

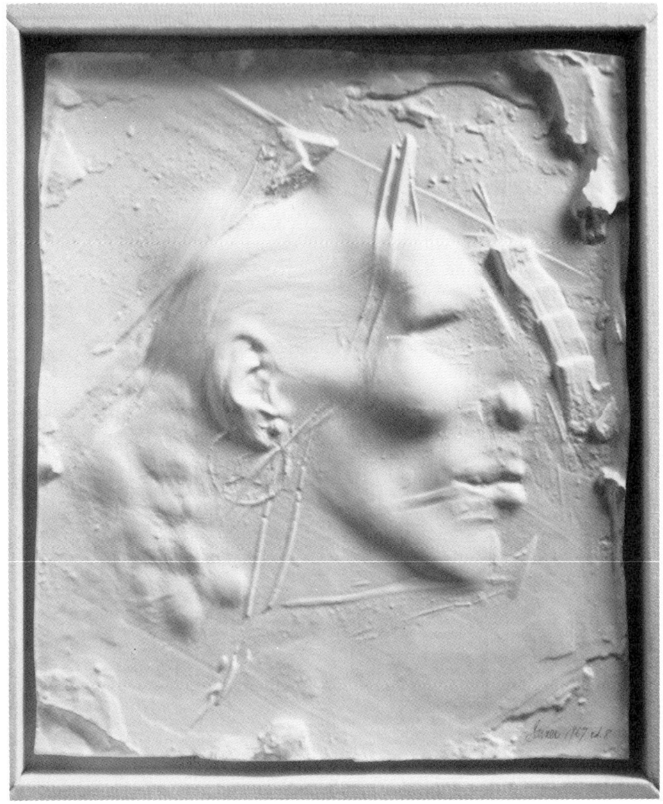

53 Tullis, *Jory I,* 1969.

shops flourish as well. Workshops such as Experimental Printmaking allow a complete mastery over image and materials.

Contemporary bookmakers also use paper in a structural way. The book may be thought of as sequential planes of image and text. Walter Hamady likens the spontaneity necessary in designing books and setting type to that needed in executing a drawing. *Gabberjab #5* [54-56] (journal liftings from November 1980) plays with letterforms, ornaments, and various sheets of paper. The work has been called by one critic "a love letter to Mary." The hundredth title issued by the Perishable Press, *Gabberjab #5* commemorates Mother's Day, 1981.

Made up of scraps of Shadwell paper left over from other editions, each page is a celebration of Gill Sans type — "love that 'g'" — and the way letter forms can be manipulated: "Just like us. You the Pear, me the sPear"; or "(5. StAnza) Clause & (break)." Pages are punched out, labels stuck on, and an account sheet in French francs stapled in. There is even an extra title

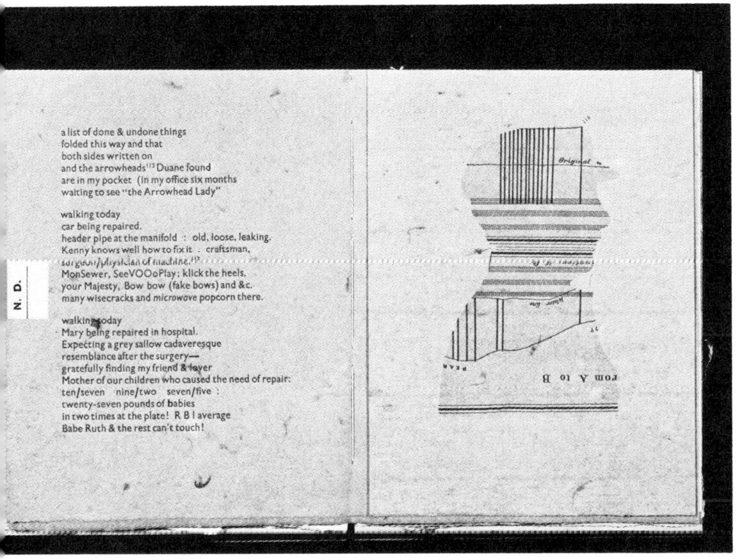

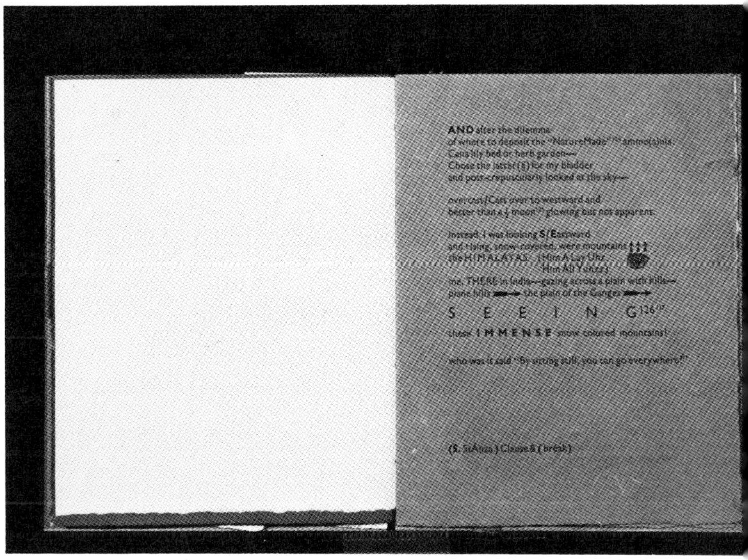

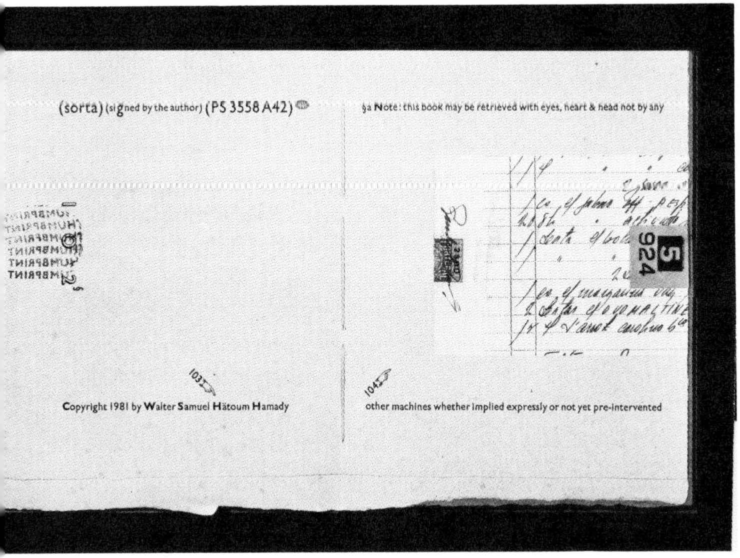

54, 55, & 56 Walter Hamady, pages from *Gabberjab #5,* 1981.

page for "librarians to do their special graffito." The book is bound with hard boards, double wrapped in twine. Footnotes, which start at number 100, are listed in a separate little book and slipped into a library pocket in the back of the book. As Hamady says in the colophon of *Gabberjab* "even an average Tin-Eye[137] can see that hands have been ON here."

Katherine Kuehn (b. 1956) of the Salient Seedling Press uses paper in a unity of image and text in her creation of *Spring Fantasy* [57]. The poet personally relates nature's springtime awakening to her own creativity. The image of the fan becomes a metaphor for the blue sky, the bursting forth of trees, flowers, and melted snow. It further symbolizes the artist's two dancing hands, which hold brush, knife, and thread. The fan ("spring captured in an abstraction") is a gift waiting to be opened.

The papers used in *Spring Fantasy*, made from "Abe's shirts, grandma's sheets, and chintzy linens from the State Street Goodwill Store," are the color of spring flowers: mauve, salmon, pink, and green. Flower fragments thrown into the pulp dot and dance in the mauve sheet. Composed of delicately colored mauve and gray pulp, the fan is subtly accented with embedded threads that peak out from underneath the mauve sheet, flower fragments, and a shimmering glitter. The fan and the book are bound with white ribbon into a full-size white sheet.

Handmade paper, pliable and sensual, may provide the "context and vehicle"[19] for explorations that extend traditional printmaking techniques. In an atmosphere of experimentation, artists are able to fuse media to solve

57 Katherine Kuehn, *Spring Fantasy*, 1978.

both painterly and sculptural problems. Whether in basement mills or in facilitating workshops, artists who create with paper work with a material that has an honorable and intimate relationship with traditional prints and drawings, a supple material that can be adapted to the needs and demands of the contemporary artist.

1. Garner Tullis, "Experimental Approaches, A Panel Discussion" in *Paper – Art & Technology* (San Francisco: World Print Council, 1979), p. 89.
2. See Janet Flint's introductory remarks in the exhibition catalog *New Ways with Paper* (Washington, D.C.: Smithsonian Institution, 1977/78).
3. See Charlotta Kotik's introductory essay in the exhibition catalog *With Paper, About Paper* (Buffalo, N.Y.: Albright-Knox Art Gallery, 1980), p. 10, for an overview of twentieth-century developments.
4. Edward B. Henning, *The Spirit of Surrealism* (Cleveland: Cleveland Museum of Art, 1979), pp. 148-49.
5. Tullis, "Experimental Approaches," p. 89.
6. Kenneth Noland, "Experimental Approaches," p. 90.
7. Ibid.
8. Richard Kubiak, *The Handmade Paper Object* (Santa Barbara, Calif.: Santa Barbara Museum of Art, 1976), unpaginated.
9. Jules Heller, "Paper as Art: A Contemporary Gallery" in *Paper – Art & Technology* (San Francisco, World Print Council, 1979), p. 97.
10. Noland, "Experimental Approaches," p. 91.
11. Kubiak, *Handmade Paper Object*.
12. Thomas Albright, "The Spirit of Santa Cruz," *Art News* (January, 1976), p. 56.
13. Paris, quoted in Jules Heller, *Papermaking* (New York: Watson-Guptill, 1978), p. 118.
14. See John Brzostoski, "Douglass Morse Howell, Scholarship, Skill, Vision," *American Craft* (February-March, 1981), p. 3.
15. Laurence Barker, "Papermaking in Spain (and America)," *Print News* (June/July, 1980), p. 4.
16. Judith Dunham, "New Forms in Paper," *Craft Horizons* (April, 1976), p. 67.
17. Albright, "Spirit of Santa Cruz," p. 54.
18. Kathy Halbreich, in a folder for the exhibition *Paper Forms* at Hayden Gallery, M.I.T., Cambridge, Mass., 1977.
19. Ibid.

Catalog

CMA = The Cleveland Museum of Art

1 Anonymous, German. *Madonna and Child with Three Members of the St. Augustine Order: Saints Augustine, Nicolas, and Clara.* Woodcut on parchment, 5 5/8 x 3 7/8 inches (14.3 x 9.9 cm.), Schreiber II.127.1057, ca. 1473. CMA, Leonard C. Hanna Jr. Collection. 28.761

2 Michel Wolgemut, German, 1434-1519. *Schatzbehälter (Reliquary).* Woodcut, 6 7/8 x 9 7/8 inches (17.5 x 25 cm.). CMA, The Harold T. Clark Educational Extension Fund. 69.519

3 Anonymous, French. *Florilegium* (folio 42). Book of forty-eight leaves; watercolor, ink, silver, and gold over occasional traces of pencil on vellum; each page approximately 12 1/4 x 8 inches (31.1 x 20.2 cm.); early seventeenth century. CMA, Gift of the American Foundation for the Maud E. and Warren H. Corning Botanical Collection. 63.594

4 Studio of Charles Le Brun (background by Adam Frans van der Meulen), French. *Louis XIV Captured by Venus.* Pencil, pen and black ink, brush and black wash on vellum, 11 1/2 x 16 1/16 inches (29.3 x 40.9 cm.), early 1670s. CMA, Delia E. Holden Fund. 69.29

5 Hans Wolff, German, died 1547. *Illustration from a Printed Book.* Woodcut with text, 3 7/8 x 6 1/8 inches (9.8 x 15.5 cm.). CMA, Gift of Mrs. Moselle Taylor Meals. 75.1023

6 Walter Hamady, American, born 1940. *Papermaking by Hand, A Book of Suspicions* (page 30). Handmade book, written by Hamady, printed by Hamady's Perishable Press on Shadwell and other handmade papers, illustrated by Jim Lee, each page approximately 10 7/8 x 7 1/4 inches (27.5 x 18.3 cm.), 1982. Lent by the artist.

7 Hamady, *Papermaking by Hand* (page 31).

8 Georges Seurat, French, 1859-1891. *Café Concert.* Conté crayon with touches of white gouache, 12 3/8 x 9 5/16 inches (31.4 x 23.7 cm.), ca. 1887. CMA, Purchase, Leonard C. Hanna Jr. Bequest. 58.344

9 Mel Pekarsky, American, born 1934. *Funeral Mountains – Study #5.* Pencil, 30 7/16 x 22 1/2 inches (77.3 x 57.2 cm.), 1979. CMA, Gift in memory of Moselle Taylor Meals. 80.100

10 Pierre Auguste Renoir, French, 1841-1919. *Baigneuse assise (Seated Bather).* Soft-ground etching in red brown, 8 5/8 x 5 5/16 inches (22 x 13.6 cm.), Delteil 11, ca. 1905. CMA, Mr. and Mrs. Charles G. Prasse Collection. 61.285

11 Käthe Kollwitz, German, 1867-1945. *Self-Portrait.* Lithograph, 10 11/16 x 9 3/16 inches (27.2 x 23.4 cm.), Klipstein 130b, 1915. CMA, Mr. and Mrs. Lewis B. Williams Collection. 41.493

12 Don Eddy, American, born 1944. *BMW Showroom.* Lithograph, 20 x 25 inches (50.8 x 63.5 cm.), 1973. CMA, The Harold T. Clark Educational Extension Fund. 74.1051

13 Robert Rauschenberg, American, born 1925. *Sunday News Magazine Cover.* Lithograph, color photo-offset, 12 1/2 x 21 1/2 inches (35.7 x 54.6 cm.), 1979. CMA, Gift of John Puskas. 81.1000

14 Charles François Daubigny, French, 1817-1878. *A Cluster of Alders (Le Bouquet d'Aunes). Cliché-verre,* 6 1/16 x 7 7/16 inches (15.3 x 18.9 cm.), Delteil 145, 1862. CMA, Gift of The Print Club of Cleveland. 25.574.

15 Clarence H. White, American, 1871-1925. *Julia Hall McCune.* Photograph, platinum print, 8 1/8 x 6 inches (20.3 x 15.2 cm.), 1903. CMA, Gift of Elizabeth Flory Kelly, John Flory, and Phoebe Flory. 80.152

16 Jerry Uelsmann, American, born 1934. *Untitled.* Photograph, silver print, 13 7/16 x 9 15/16 inches (34.1 x 25.2 cm.), 1964. Anonymous loan.

17 Robert Trostle, American, born 1941. *Geometrics 1 – Roller Coaster, Mexico City*. Photograph, cibachrome color, 20 x 16 inches (50.8 x 40.6 cm.), 1978. CMA, Gift of Joseph Erdelac. 78.1023

18 Charles Bell, American, born 1935. *Pin Ball*. Photograph, cibachrome color, 7 7/16 x 9 5/8 inches (19 x 24.4 cm.), 1981. CMA, Gallery Group Fund. 81.1008

19 Hilda Volkin, American, born 1933. *Figure I*. Screen print in colors on Oriental paper, 22 x 26 inches (55.9 x 66 cm.), 1973/74. CMA, Gift of Hilda Volkin. 74.260

20 Ernst Ludwig Kirchner, German, 1880-1938. *Dune Landscape with a Curving Bay*. Woodcut on blotting paper, 17 x 13 3/4 inches (43.1 x 35 cm.), Dube 220 a/b, 1913. CMA, Gift of Moselle Taylor Meals. 73.235

21 Claire Falkenstein, American, contemporary. *Untitled*. Etching, embossed, 21 3/16 x 17 5/16 inches (53.8 x 44 cm.), 1969. CMA, The Harold T. Clark Educational Extension Fund. 69.543

22 Victor Vasarely, French (b. Hungary), born 1908. *G-FF*. Screen print, embossed, 10 1/4 x 9 3/16 inches (26.1 x 25 cm.), 1967. CMA, Gift of The Print Club of Cleveland. 68.274

23 Paolo Boni, Italian, born 1926. *Extase*. Etching, embossed, 5 7/8 x 6 7/8 inches (14.6 x 17.5 cm.), 1967. CMA, Gift of Mr. and Mrs. Wilbur D. Prescott. 77.1027

24 Paolo Boni. Metal plate for *Extase*. 5 7/8 x 6 7/8 inches (14.6 x 17.5 cm.), 1967.

25 Emile Nolde, German, 1867-1956. *Dancers (Tänzerinnen)*. Woodcut, 9 3/8 x 12 1/4 inches (23.8 x 31.2 cm.), Schiefler and Mosel II.132, 1917. CMA, Delia E. Holden Fund. 60.158

26 Louise Nevelson, American (b. Russia), born 1899. *Beyond-Q 17*. India ink with collage, 30 1/4 x 22 inches (76.9 x 56.2 cm.), 1977. CMA, Gift of The Print Club of Cleveland. 77.96

27 Ellen Stavitsky, American, born 1931. *Number 16 ("Visual Delights" Series)*. Collage, 15 x 11 inches (38.1 x 11.1 cm.), 1979. CMA, The Harold T. Clark Educational Extension Fund. 81.1019

28 Frank Stella, American, born 1936. *Untitled*. Lithograph printed in colors, pencil added by hand, 17 1/16 x 22 inches (43.3 x 55.9 cm.), 1967. CMA, Anonymous Gift for the Fiftieth Anniversary of The Print Club of Cleveland. 70.25

29 Ed Moses, American, born 1926. *Wedge 5*. Lithograph printed in colors, on three sheets of tissue paper, 23 5/8 x 17 7/8 inches (60 x 45.4 cm.), 1973. CMA, Gift of Mr. and Mrs. Joseph T. Meals. 79.128

30 Alan Shields, American, born 1944. *Sun Moon Title Page*. Double-sided collage with screen printing, monoprinting, watercolor, and sewing, 26 1/4 x 26 1/4 inches (66.6 x 66.6 cm.), 1971. CMA, Gift of Mr. and Mrs. Robert M. Levine. 74.191

31a George Bellows, American, 1882-1925. *The Black Hat*. Lithograph on white paper, 13 1/16 x 9 3/16 inches (33.1 x 23.1 cm.), Mason 113, 1921. CMA, Leonard C. Hanna Jr. Collection. 37.484

31b George Bellows. *The Black Hat*. Lithograph on cream paper, 13 1/16 x 9 3/16 inches (33.1 x 23.1 cm.), Mason 113, 1921. CMA, Mary Spedding Milliken Memorial Collection, Gift of William Mathewson Milliken. 50.406

32 Lorenzo di Credi, Italian, 1459/60-1537. *Madonna and Child*. Silverpoint on salmon prepared paper, 5 11/16 x 3 11/16 inches (14.5 x 9.4 cm.), ca. 1510. CMA, John L. Severance Fund. 63.472

33 Ludovico Cardi (called Cigoli), Italian, 1559-1613. *Standing Male Figure*. Brush and brown ink, heightened with white on paper prepared with green wash, 15 13/16 x 10 7/8 inches (40.2 x 27.6 cm.), ca. 1593. CMA, The Charles W. Harkness Endowment Fund. 29.555

34 Jean-Baptiste Oudry, French, 1689-1755. *The Fox and the Turkey Hens.* Pen and black and gray ink, brush and gray wash, heightened with white gouache, blue watercolor border, on blue paper, 12¼ x 10¼ inches (31.2 x 26.1 cm.), 1733. CMA, John L. Severance Fund. 77.93

35 Camille Pissarro, French, 1830-1903. *Nude Women (Femmes Nues).* Lithograph on cream paper, 8$^{5}/_{16}$ x 11$^{1}/_{16}$ inches (21.1 x 28.1 cm.), Delteil 157, design ca. 1894, printed 1922. CMA, Gift of The Print Club of Cleveland. 73.124

36 Camille Pissarro. *Nude Women (Femmes Nues).* Lithograph on pink *chine collée,* 8$^{5}/_{16}$ x 11$^{1}/_{16}$ inches (21.1 x 28.1 cm.), design ca. 1894, printed 1923. CMA, Gift of The Print Club of Cleveland. 73.125

37 Walter Hamady. *Making a Sacher Torte* (frontispiece). Handmade book, written by Diane Wakoski, printed by Hamady's Perishable Press on Shadwell handmade paper, illustrated by Ellen Lanyon, each page approximately 11⅛ x 7⅜ inches (28.2 x 18.1 cm.), 1981. Lent by the artist.

38 Walter Hamady. *Making a Sacher Torte* ("Sally Plum" poem).

39 William Morris, English, 1834-1896. *News from Nowhere or an Epoch of Rest.* Handmade book, written by Morris, published by the Kelmscott Press, on handmade paper, each page approximately 8$^{3}/_{16}$ x 5½ inches (20.5 x 14 cm.), 1892. CMA, Bequest of Julia Morgan Marlatt.

40 Félix Hilaire Buhot, French, 1847-1898. *The Country Neighbors.* Etching, drypoint, and aquatint, 5⅜ x 7⅛ inches (13.6 x 18.1 cm.), Bourcard and Goodfriend 148, state III/V. CMA, Gift of Mrs. Malcolm L. McBride. 65.259

41 Félix Buhot. *The Country Neighbors.* Etching, drypoint, and aquatint, 5¼ x 7$^{1}/_{16}$ inches (13.4 x 17.9 cm.), Bourcard and Goodfriend 148, state V/V. CMA, Leonard C. Hanna Jr. Collection. 47.455

42 James McNeill Whistler, American, 1834-1903. *The Garden.* Etching printed in brown ink, 12 x 9¼ inches (30.4 x 23.5 cm.), Kennedy 210, state V/VIII, 1886. CMA, Gift of Mr. and Mrs. Ralph King. 23.31

43 Jasper Johns, American, born 1930. *Numerals 0-9.* Etching and aquatint, 17⅛ x 13$^{11}/_{16}$ inches (43.5 x 34.7 cm.), 1975. CMA, Gift of The Print Club of Cleveland. 76.133

44 Inez Storer, American, born 1933. *Wrapped 1.* Paper piece, 28 x 22 inches (71.1 x 55.8 cm.), 1976. CMA, The Harold T. Clark Educational Extension Fund. 81.1014

45 William Weege III, American, born 1935. *Oconomowoc.* Paper piece, 21 x 26 inches (53.3 x 66.1 cm.), 1979. CMA, The Harold T. Clark Educational Extension Fund. 80.1037

46 Suzanne Anker, American, born 1946. *Ribbed Series, #3.* Paper piece, 28½ x 22 inches (72.4 x 55.8 cm.), 1975. CMA, The Harold T. Clark Educational Extension Fund. 81.1015

47 Richard Hunt, American, born 1935. *Paper Piece II (Untitled).* Paper piece, 16¼ x 21 inches (41.3 x 53.3 cm.), 1978. CMA, The Harold T. Clark Educational Extension Fund. 79.1029

48 Harold Paris, American, 1925-1979. *Carta Series, #50.* Paper piece, 11 x 11½ inches (27.9 x 29.2 cm.), 1978. CMA, The Harold T. Clark Educational Extension Fund. 81.1013

49 June Bonner, American, born 1918. *Untitled Paper 2.* Paper piece, 8¼ x 8 inches (21 x 20.3 cm.), 1979. CMA, The Harold T. Clark Educational Extension Fund. 81.1024

50 Laurence Barker, American, born 1930. *Untitled #5.* Paper piece, 14$^{9}/_{16}$ x 18$^{5}/_{16}$ inches (37 x 46.5 cm.). CMA, In Memory of Janet Mack (Sundry Purchase Art). 80.196

51 Claire Falkenstein, American, contemporary. *Never Ending Screen.* Paper piece, diam. 19¼ inches (48.9 cm.), 1975. CMA, The Harold T. Clark Educational Extension Fund. 81.1012

52 Garner Tullis, American, born 1939. *Technological Artifact – Indian.* Paper piece, 23 x 20 inches (58.4 x 50.8 cm.), 1975. CMA, Gift of the artist. 76.119

53 Garner Tullis. *Jory I.* Vacuum-pressed plastic, 20½ x 18 inches (52 x 45.7 cm.), 1969. CMA, Anonymous Gift. 71.222

54 Walter Hamady, American, born 1940. *Gabberjab #5* (page headed "a list of done & undone things"). Handmade book, written by Hamady, printed by Hamady's Perishable Press on Shadwell handmade paper, each page approximately 6⅞ x 5⅜ inches (17.3 x 13.2 cm.), 1981. Lent by the Artist.

55 Hamady, *Gabberjab #5* (page headed "And after the dilemma").

56 Hamady, *Gabberjab #5* (copyright page).

57 Katherine Kuehn, American, born 1956. *Spring Fantasy.* Book and paper piece, written by Kuehn, printed by Kuehn's Salient Seedling Press on Palm Box handmade paper, 21 x 14⅛ inches (53½ x 36 cm.), 1978. Lent by the artist.

Glossary

Beating Process of macerating rag or other material with water into pulp.

Calendering Process to smooth or gloss paper by either passing it between rollers or placing it between metal plates under pressure.

Casein An adhesive, used to coat paper, made of protein obtained from skimmed milk.

Cast paper A method of creating a paper piece by pouring pulp into a mold or by patting pulp into a form.

Cellulose A carbohydrate and main component of the cell walls of plants; it constitutes the long, hollow chains called fibers. These fibers will adhere together to form a mat from a water suspension, e.g., paper.

Chiaroscuro In drawing, the graphic arts, and painting, the rendering of form through dark and light areas, so as to achieve an illusion of depth.

Cliché-verre A photographic print method in which a design is scratched onto a glass plate coated with collodion; this plate is then put on light-sensitive paper and exposed to sunlight. Where the collodion was scratched out, the paper turns black.

Couch Transfer of the layer of wet pulp from the mold to a dampened felt.

Daguerreotype A photographic method (invented by 1839 by Niepce and Daguerre) that forms an image on a metal plate in a camera.

Deckle The wooden frame or fence that fits on the mold during sheet formation to contain the pulp.

Deckle edge The feathered edges on a sheet of paper caused by the deckle.

Dye Soluble coloring matters that impart their colors to other substances by staining or being imbibed by them.

Embossment Process of creating raised figures or designs in relief.

Felt A woven cloth, usually wool, onto which a newly formed sheet is couched; also, removing water from the sheet with a sponge.

Finishing Drying, sizing, and calendering sheets of paper.

Gampi One of the three plants most commonly used in Japanese papermaking; it grows only in the wild and is less absorbent than other fibers.

Gelatin One of many sizings for paper made by heating scraps from animal hides.

Half-tone screen process A method that breaks down an image into a dot pattern that reproduces a varying range of grays onto a printing plate.

Hemp An Asian plant high in cellulose used for papermaking; one of the oldest kinds of true paper.

Hog A mechanical agitator for stirring pulp in the vat.

Hollander beater A machine invented in Holland by 1680 to break down rag fibers in the preparation of pulp.

Incunabula Literally "out of the cradle," the term most frequently refers to books produced during the first fifty years of typographic printing: 1450-1500.

Intaglio A print process in which the design is carved or etched into the surface of a plate; the plate is inked, the surface wiped clean, and the ink is transferred to damp paper by running plate and paper through a press.

Kozo One of the three principal fibers used in Japanese papermaking; made from mulberry plants, it generally makes the strongest of the Japanese papers.

Laid mold A wooden frame for forming pulp into sheets; vertical (laid) metal wires are held in place by horizontal chain wires supported by tapered wooden ribs in a wooden frame.

Laid paper Paper formed on a laid mold; the impression made in the sheet during formation by the vertical and horizontal lines on the mold is visible in a grid pattern in the sheet.

Layer The person who separates each sheet of paper from the interleaved felts after the post has been pressed.

Lithography Planographic or "flat writing"; a printing method in which a design is made on a stone with a greasy crayon. After the stone is dampened, ink is rolled across its surface; the ink, which sticks to the greasy crayon marks and not to the stone, is transferred to paper in a press.

Litho transfer paper Special paper on which the artist draws a design, which can then be transferred to a litho stone.

Mitsumata One of the three main plants used in Japanese papermaking; its fibers are soft and absorbent.

Mold, or mould The papermaker's basic tool, used to make the sheet. Dipped into the vat, the mold forms a mat of fibers that drains and is couched onto felts.

Neri Mucilage or vegetable starch extracted from roots or bark of trees; it looks like egg white and works to make the fibers float evenly in the vat without clumping, to keep fibers from entangling during formation, and to prevent each sheet from sticking to the next after the stack is pressed.

Offset lithography The transfer of an inked impression from a stone to a rubber roller that picks up the impression and deposits it on paper; widely used in commercial printing.

Paper A felted or weblike mass of interlaced plant fibers in sheet form; produced by flowing a dilute slurry of fibers in water across a screen; the water is allowed to drain; the sheet is then removed from the screen, pressed, and dried.

Parchment A writing or printing surface made from the skins of animals, often sheep or goats.

Pigment A coloring substance insoluble in the liquid vehicle with which it is mixed; it imparts its color effect by being spread over a surface.

Plate mark An indentation in a sheet of paper created by the edge of a plate during intaglio printing.

Post A pile of wet sheets separated by felts.

Prepared paper Paper coated with a mixture of ground bone or eggshell, pigment, and a binder.

Pulp The product of the beating process in papermaking.

Relief printing A printing method in which the image is printed from the surface of the block or plate with all non-printing areas carved, engraved, or etched away; remaining lines and areas are inked and printed by hand or in a press.

Remarques Sketches surrounding the main image of a print.

Retting Leaving rags to rot in piles so that resins in the fibers weaken in preparation for the stamping mill.

Screen print or serigraph Made from a stencil adhered to a woven screen that blocks out selected areas; the screen is placed on a sheet of paper and ink is forced through the open spaces in the woven mesh with a squeegee.

Signature A section of a book made of a folded sheet.

Sizing A substance added to paper to control ink or paint absorption.

Soft-ground etching An intaglio printing method in which the metal plate is prepared with a soft ground that does not completely harden; textures pressed onto the ground make a pattern, which can then be etched with acid and printed.

Stamper The machine used in the original method of beating rags and other materials into pulp; it works like a mortar and pestle.

State A change in a printing plate.

Stuff Pulp ready for the vat.

Tripan Platform that extends across the vat that allows water to drain from a newly formed sheet while on the mold.

Vat A tub that holds pulp into which the mold is dipped to form a sheet.

Vat person The person who forms the sheet by dipping a mold into the vat and then shaking the mold to each side and to and from the body to set the fibers and form the sheet.

Vellum A writing or printing surface made from the skin of animals; a refined parchment.

Watermark A thinner area in the sheet caused by wire sewn to the face of the mold, which leaves its imprint in the wet pulp; it may be the logo or trademark of a mill or paper-maker.

Wove mold A mold made from a woven wire screen stretched across a wooden frame supported by tapered wooden ribs.

Wove paper A sheet made on a wove mold; it has a finer, smoother surface than paper formed on a laid mold.

Selected Bibliography

Adams, Alice. "Douglass Howell." *Craft Horizons* 22 (1962): 26-29.

Albright, Thomas. "The Spirit of Santa Cruz." *Art News* 75 (1976): 52-56.

Ames-Lewis, Francis. *Drawing in Early Renaissance Italy.* New Haven and London: Yale University Press, 1981.

Barker, Laurence. "Papermaking in Spain (and America)." *Print News* 2, no. 3 (1980): 2-6.

Barnard, Osbert H. "The Clichés-Verres of the Barbizon School." *The Print Collector's Quarterly* 9 (1922): 149-72.

Barrett, Timothy. *Nagashizuki – The Japanese Craft of Hand Papermaking.* North Hills, Pa.: Bird and Bull Press, 1979.

Blum, André. *On the Origin of Paper.* Translated by H. M. Lydenberg. New York: R. R. Bowker, 1934.

Brzostoski, John. "Douglass Morse Howell, Scholarship, Skill, Vision." *American Craft* 41, no. 1 (1981): 2-5.

Dunham, Judith L. "New Forms in Paper." *Craft Horizons* 36 (1976): 38-43.

Eichenberg, Fritz. *The Art of the Print: Masterpieces, History, Techniques.* New York: Harry N. Abrams, 1976.

Flint, Janet. *New Ways with Paper* [exh. cat.]. Washington, D.C.: Smithsonian Institution, 1977/78.

Green, J. Barcham. *One Hundred and Fifty Years of Papermaking by Hand.* Maidstone, England: J. Barcham Green, 1960.

Halbreich, Kathy. *Paper Forms* [exh. folder]. Cambridge, Mass.: Hayden Gallery, M.I.T., 1977.

Hamady, Walter. *Papermaking by Hand, A Book of Suspicions.* Mt. Horeb, Wis.: Perishable Press, 1982.

Heller, Jules. *Papermaking.* New York: Watson-Guptill, 1978.

Henning, Edward B. *The Spirit of Surrealism* [exh. cat.]. Cleveland: Cleveland Museum of Art, 1979.

Howell, Douglass. "The Future of Paper." *Print* 14 (1960): 95-96.
———. "Papers for Printmaking." *Art News* 49 (1963): 63.
Hughes, Sukey. *Washi, The World of Japanese Paper.* Tokyo, New York, and San Francisco: Kodansha International, 1978.
Hunter, Dard. *A Papermaking Pilgrimage to Japan, Korea and China.* New York: Pynson Printers, 1936.
———. *Papermaking: The History and Technique of an Ancient Craft.* 2d ed. rev. & enl. New York: Alfred A. Knopf, 1947.
———. *Papermaking by Hand in America.* Chillicothe, Ohio: Mountain House Press, 1950.
———. *My Life With Paper.* New York: Alfred A. Knopf, 1958.
Imberdis, J. *Papyrus sive Ars conficeindae Papyri [Papyrus or the Craft of Paper*]. 1693. Reprint. North Hills, Pa.: Bird and Bull Press, 1961.
Johnson, Fridolf. "Henry Morris, Papermaker and Printer." *American Artist* 31 (1967): 56-61.
Kotik, Charlotta. *With Paper, About Paper* [exh. cat.]. Buffalo, N.Y.: Albright-Knox Art Gallery, 1980.
Kubiak, Richard *The Handmade Paper Object* [exh. cat.]. Santa Barbara, Cal.: Santa Barbara Museum of Art, 1976.
Kunisaki, Jihei. *Kamisuki Chohoki; A Handy Guide to Papermaking.* After Jap. ed. of 1798. Translated by Charles E. Hamilton. Berkeley: University of California Press, 1948.
Lewis, Golda. "Douglass Howell: Sketchbooks." *Craft Horizons* 36 (1976): 30-31.
Long, Paulette, ed. *Paper – Art and Technology.* San Francisco: World Print Council, 1979.
Lucie-Smith, Edward. *The Story of Craft, The Craftsman's Role in Society.* Ithaca, N.Y.: Cornell/Phaidon Books, 1981.
Maberly, Joseph. *The Print Collector.* 2d. ed. New York, 1880.
Meder, Joseph. *The Mastery of Drawing.* Vol. 1. Translated and revised by Winslow Ames. New York: Abaris Books, 1978.
Morris, Henry. *The Paper-Maker.* North Hills, Pa.: Bird and Bull Press, 1974.
Papermaking: Art and Craft [exh. cat.]. Washington, D.C.: Library of Congress, 1968.
Robison, Andrew. *Paper in Prints* [exh. cat.]. Washington, D.C.: National Gallery of Art, 1977.
Rose, Barbara. "The News Is Paper." *Vogue.* December, 1974.
Schiller, Marlene. "The American Community of Hand Papermakers." *American Artist* 41, no. 421 (1977): 39-44.
Schlosser, Leonard. "The Graphic Confluence of 1800." In *A Miscellany for Bibliophiles,* edited by H. George Fletcher. New York: Grastorf and Lang, 1981.
Studley, Vance. *The Art and Craft of Handmade Paper.* New York: Van Nostrand Reinhold, 1977.
Thompson, Susan Otis. *American Book Design and William Morris.* New York and London: R. R. Bowker, 1977.
Watrous, James. *The Craft of Old Master Drawings.* Madison, Wis.: University of Wisconsin Press, 1957.
Young, Joseph E. "Pages and Fuses: An Extended View of Robert Rauschenberg." *Print Collector's Newsletter* 5 (1974): 25-29.